Cartography Design Annual

#1

Edited by Nick Springer

For information contact Nick Springer, PO Box 72, Crosswicks, NJ, 08515.

ISBN-13: 978-0-6152-2116-8

Second Edition, August 2008

Designed by Nick Springer.

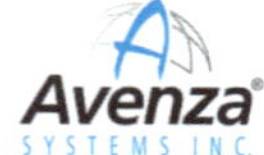

Financial support provided by Avenza Systems Inc., Mapping Solutions for the 21st Century®
www.avenza.com

Printed in USA by Lulu.com

Contents

Foreword

While I had studied some aspects of technical cartography in pursuit of my degree in geography, I was more enamored of the design classes and my hands-on work in the cartography lab. After graduation I fell into a career in graphic design because I never found any cartographic design jobs. I still created maps here and there but that was only an occasional fun project. Because of this I never really became involved in any kind of cartographic community.

Years later I finally got a full time job as a cartographic designer, and eventually established a successful freelance business. As my daily involvement with cartography grew I began to feel isolated as a cartographer and wondered how to connect with other practitioners. I had never come into contact with anyone else and thought that maybe I was one of the few. I searched the web in vain for any kind of online community for cartographers, and so in April of 2004 I decided to create Cartotalk.com in hopes of creating a community on my own.

Cartotalk started slowly, not gaining any real momentum until 2005 at which time I also learned about the North American Cartographic Information Society (NACIS). I realized that there were plenty of cartographers around, and through lots of comments on Cartotalk, it seemed like we all yearned for a greater sense of community.

I am very proud that Cartotalk has turned out to be such a vibrant place for cartographers to share ideas, get peer feedback, ask for and receive advice, and generally just communicate. I have seen a lot of great maps posted on Cartotalk and it occurred to me that the best of these should be compiled and shared, thus the impetus for this book.

The goal was not to have a competition in the sense of judging, but to try to select the best designed maps of the year, while being as inclusive as possible. The process was agnostic towards the tools used to create the maps and indeed some of the maps were created wholly in GIS systems, most used graphic design tools, and some were hand-drawn.

Clearly this book is not a comprehensive collection since submissions were solicited mostly through Cartotalk and it's highly likely that many skilled cartographers did not know about the book for this first edition. Because of this, and because it was not meant to be competitive, it is called an "Annual" and not a "Best of" collection.

Hopefully promotion of this first book, and word of mouth, will make next year's annual an even better representation of the breadth of excellent cartographic design work that is taking place now in 2008. The maps in this edition were required to have been first published in 2007, although some copyrights are listed as 2008 due to republishing or second editions. Most of the maps chosen to be included were submitted in response to the call for submissions, but a few were maps that I had seen during the last year and requested the author submit them. In 2008 I will be on the lookout for more maps, and will be asking for submissions starting in January of 2009.

The process for selecting the maps to include was purely my own opinion. I based it on the aesthetic value, the clarity of the data, and unique aspects of the design. I selected two maps from a few cartographers where I thought that there was a divergent enough style that the works did not seem to be redundant. And yes, I did include some of my own work.

In the end, my real goal is to simply showcase beautiful maps. I would love to hear that this book has caught the interest of people outside the circle of cartographers and helps us show people that the *art* of map making is still very much alive.

Enjoy.

Nick Springer

Hugo Ahlenius, UNEP/GRID-Arendal

The Cryosphere, World Map

Software used: ESRI ArcGIS, Adobe Illustrator

Data sources: National Snow and Ice Data Center (NSIDC), International Permafrost Association (IPA), VMAP

More information: maps.grida.no/go/graphic/the-cryosphere-world-map

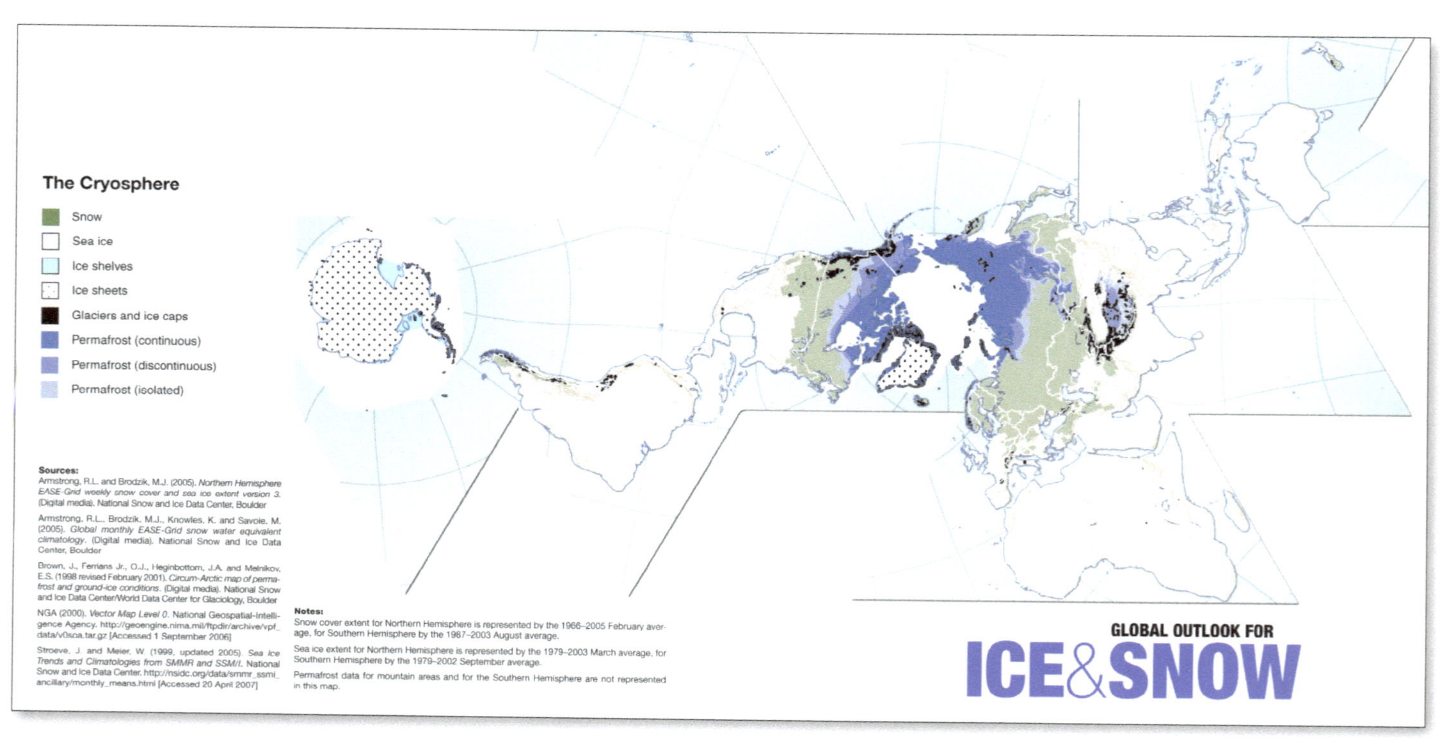

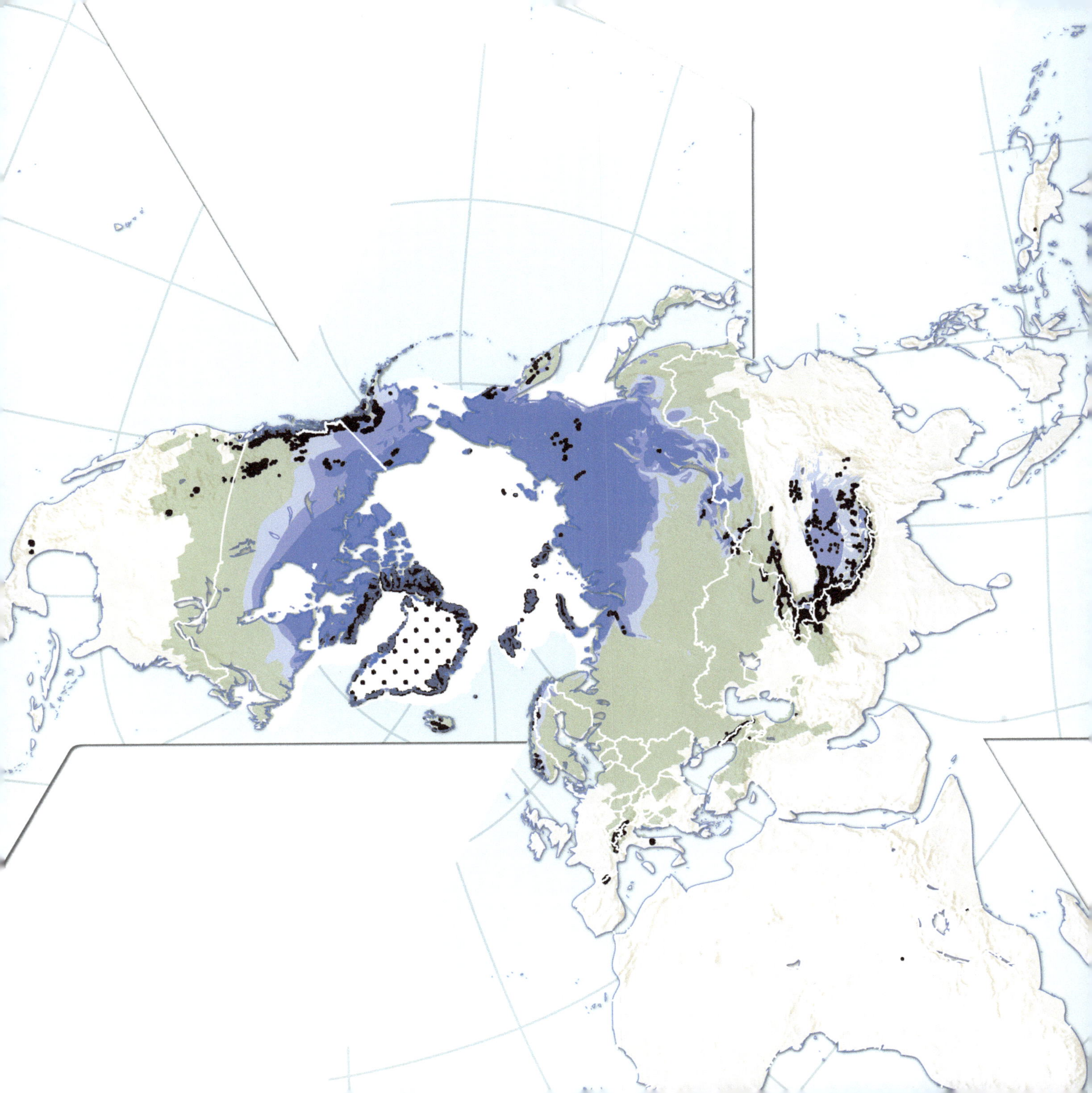

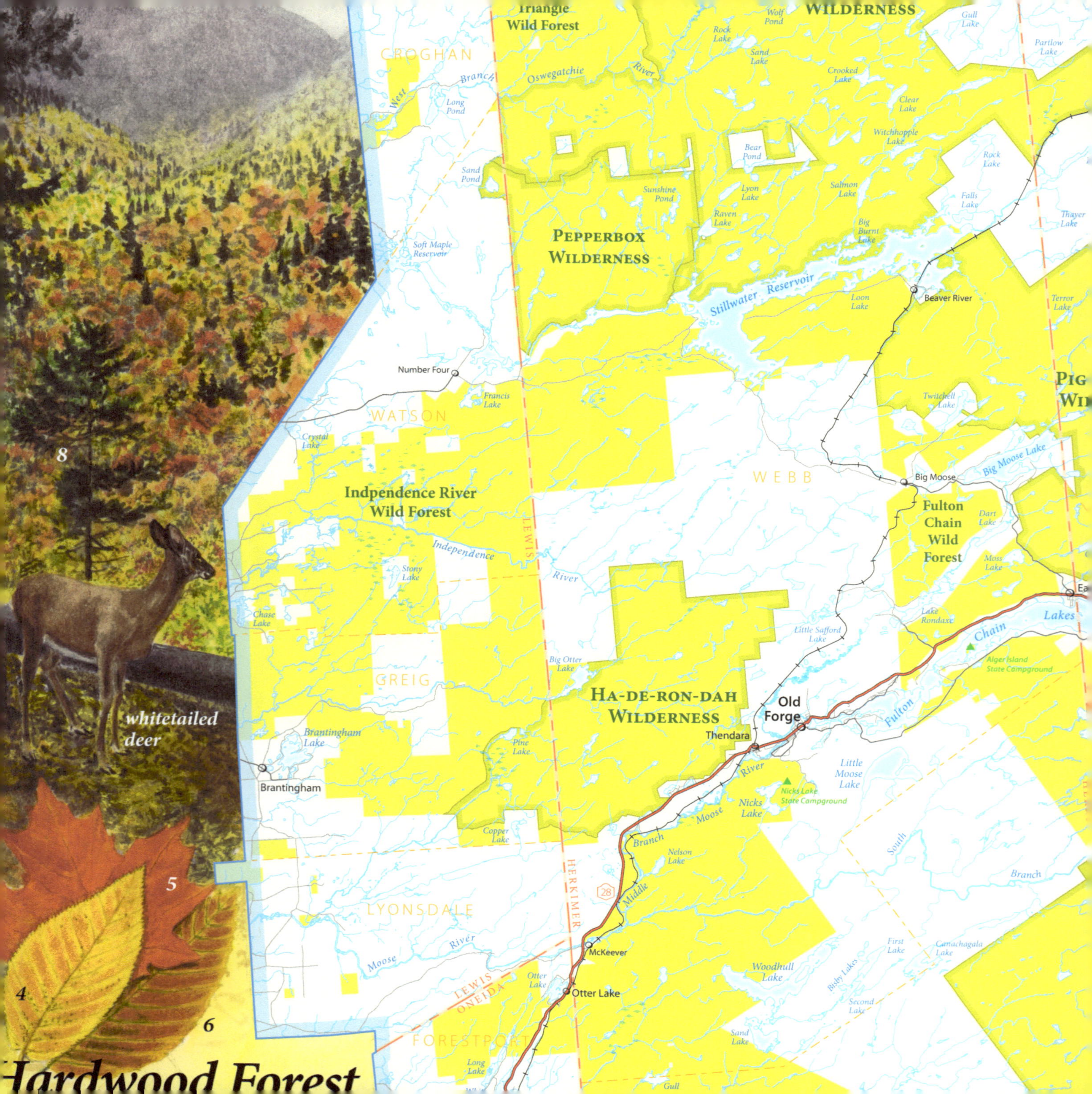

Triangle Wild Forest
WILDERNESS
CROGHAN
West Branch
Oswegatchie River
Wolf Pond
Rock Lake
Sand Lake
Gull Lake
Partlow Lake
Crooked Lake
Clear Lake
Long Pond
Witchhopple Lake
Rock Lake
Bear Pond
Sand Pond
Sunshine Pond
Lyon Lake
Salmon Lake
Falls Lake
Raven Lake
Thayer Lake
Big Burnt Lake
PEPPERBOX WILDERNESS
Soft Maple Reservoir
Stillwater Reservoir
Loon Lake
Beaver River
Terror Lake
Number Four
Francis Lake
WATSON
Twitchell Lake
PIG
WI
Crystal Lake
8
Big Moose Lake
Big Moose
WEBB
Indpendence River Wild Forest
Fulton Chain Wild Forest
Dart Lake
Independence River
LEWIS
Stony Lake
Moss Lake
Chase Lake
Lake Rondaxe
Little Safford Lake
Lakes
Chain
Alger Island State Campground
Big Otter Lake
GREIG
HA-DE-RON-DAH WILDERNESS
Old Forge
whitetailed deer
Thendara
Fulton
Brantingham Lake
Pine Lake
Brantingham
Little Moose Lake
Moose River
Nicks Lake State Campground
Nicks Lake
Copper Lake
Branch
Nelson Lake
South Branch
5
HERKIMER
28
Middle
LYONSDALE
Moose River
McKeever
First Lake
Canachagala Lake
Woodhull Lake
Bisby Lakes
Otter Lake
Otter Lake
LEWIS
ONEIDA
4
Second Lake
6
Sand Lake
FORESTPORT
Long Lake
Gull
Hardwood Forest

John W. Barge, MapRelief

The Adirondack Park – A Park of People and Natural Wonder

Illustrations by Anne E. Lacy

Software used: ESRI ArcGIS, Adobe Illustrator, Avenza MAPublisher

More information: www.adirondackcouncil.org or MapRelief@gmail.com

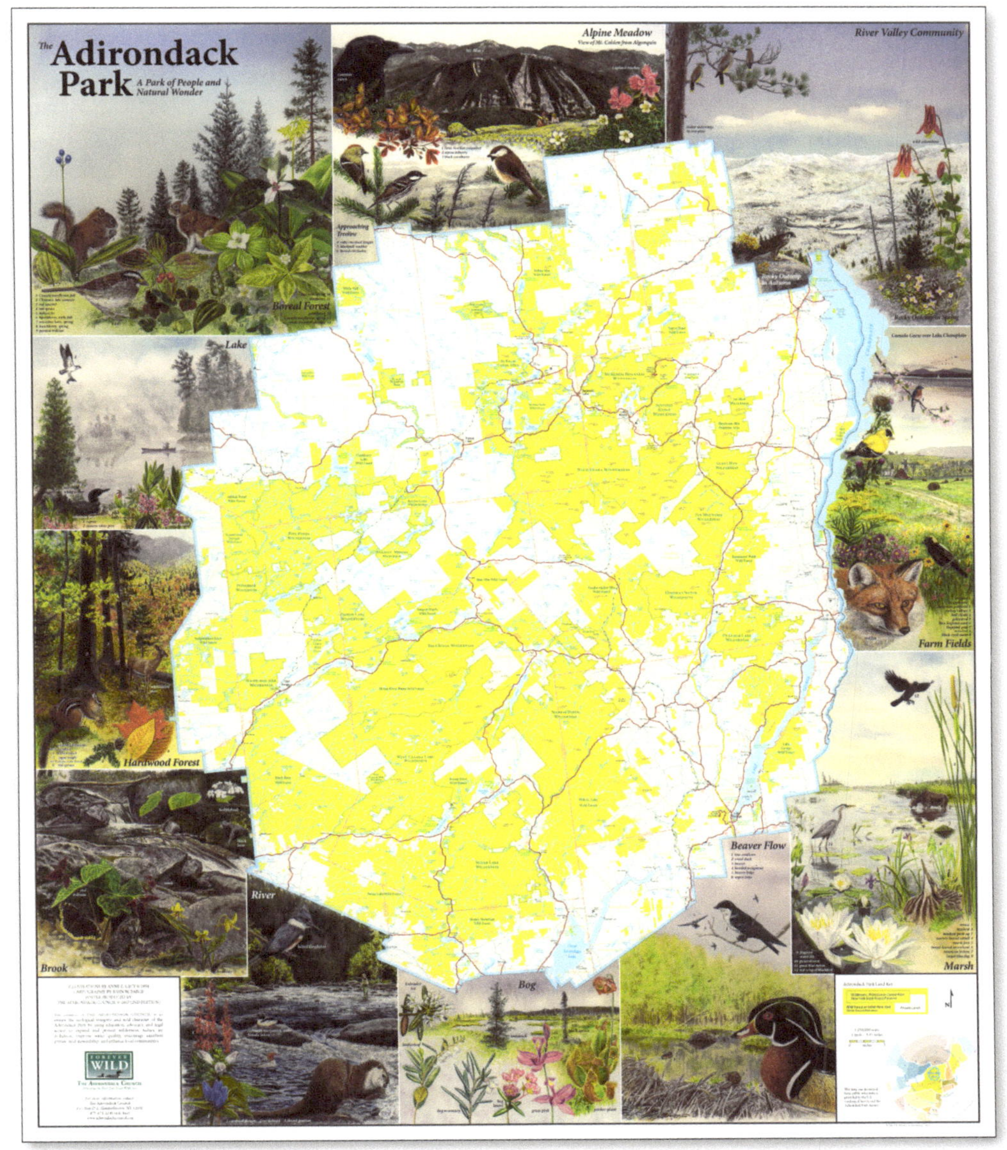

Benchmark Maps

Wyoming Road & Recreation Atlas

1st Edition

Software used: Macromedia FreeHand, ESRI ArcMap, Adobe Photoshop, FileMaker, Quark XPress

Data sources: Data sets originate as publicly available data and then are compiled from other sources via desk-editing and from months of field-checking by driving the land and interviewing the local experts.

More information: www.benchmarkmaps.com

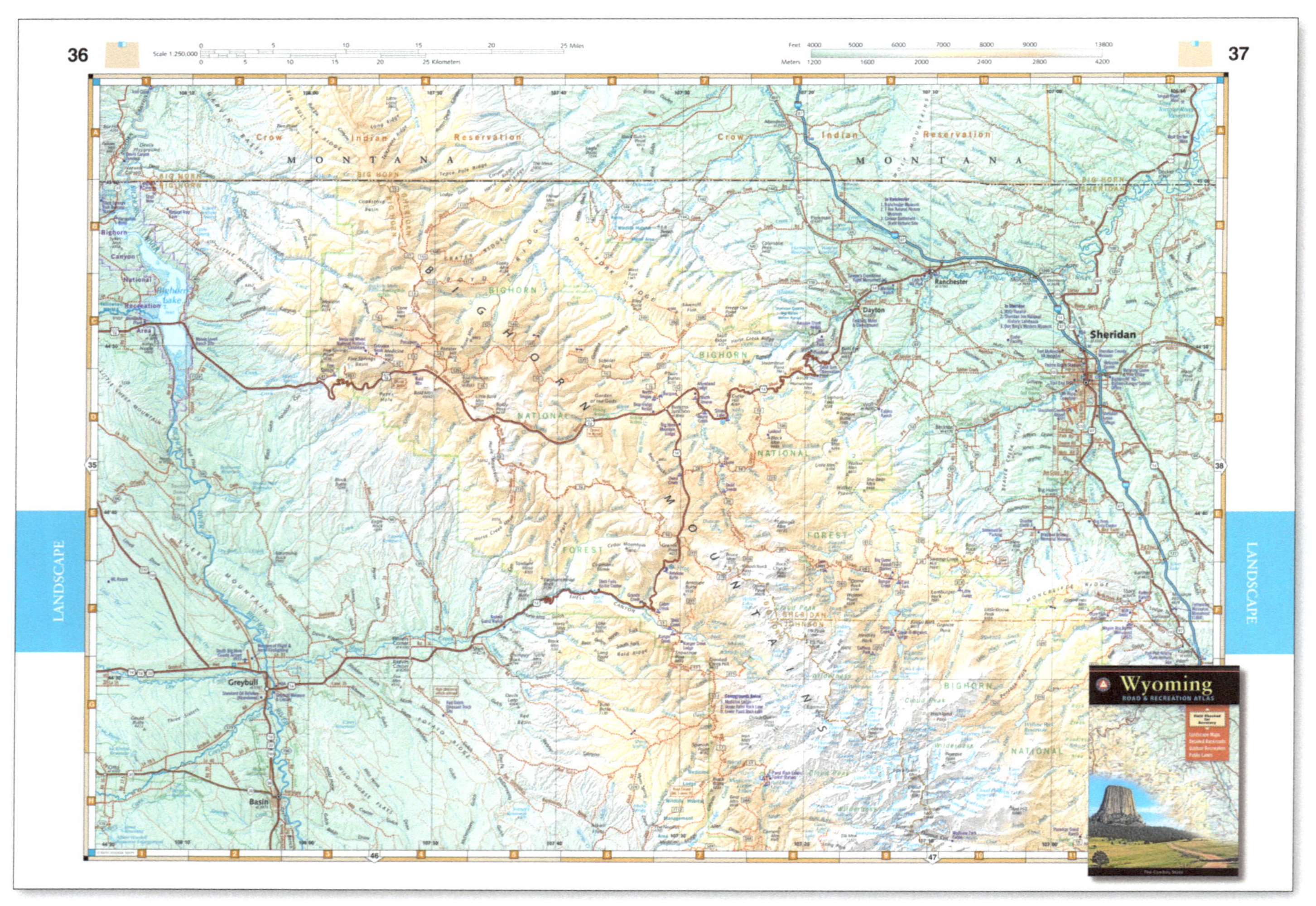

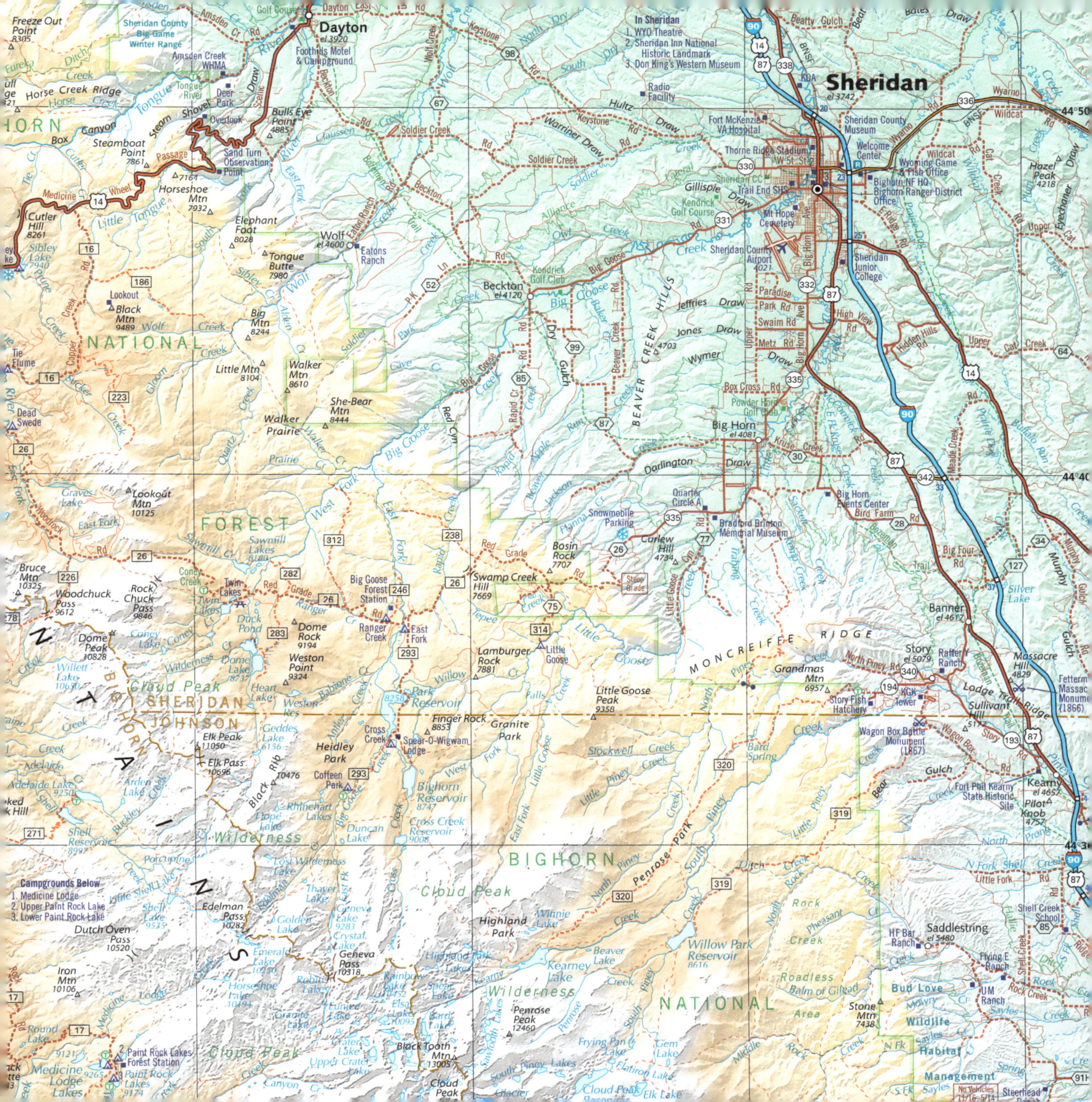

Sheridan
el 3742
In Sheridan
1. WYO Theatre
2. Sheridan Inn National Historic Landmark
3. Don King's Western Museum
Dayton
el 3920
Foothills Motel & Campground
Sheridan County Big Game Winter Range
Amsden Creek WHMA
Freeze Out Point
8305
Horse Creek Ridge
Tongue River
Deer Park
Steamboat Point
7861
Bulls Eye Point
4885
Overlook
Sand Turn Observation Point
Horseshoe Mtn
7932
Cutler Hill
8261
Sibley Lake
Elephant Foot
8028
Wolf
el 4600
Eatons Ranch
Tongue Butte
7980
Lookout
Black Mtn
9489
Big Mtn
8244
Little Mtn
8104
Walker Mtn
8610
She-Bear Mtn
8444
Walker Prairie
Tie Flume
Dead Swede
NATIONAL
FOREST
Lookout Mtn
10125
Bruce Mtn
10325
Woodchuck Pass
9612
Rock Chuck Pass
9846
Twin Lakes
Sawmill Lakes
Big Goose Forest Station
Ranger Creek
East Fork
Dome Peak
10828
Dome Rock
9194
Weston Point
9324
Willett Lake
Cloud Peak
SHERIDAN
JOHNSON
Elk Peak
11050
Elk Pass
10696
Heidley Park
Coffeen Park
Cross Creek
Spear-O-Wigwam Lodge
Finger Rock
8853
Granite Park
Park Reservoir
Bighorn Reservoir
8747
Cross Creek Reservoir
9008
Duncan Lake
Adelaide Lake
9250
Arden Lake
Shell Reservoir
Porcupine
Lost Wilderness Lake
Thayer Lake
Geneva Lake
9283
Crystal Lake
Geneva Pass
10318
Golden Lake
Edelman Pass
10282
Dutch Oven Pass
10520
Iron Mtn
10106
Emerald Lake
Horseshoe Lake
Round Lake
Medicine Lodge Lakes
Paint Rock Lakes Forest Station
Paint Rock Lakes
9174
Campgrounds Below
1. Medicine Lodge
2. Upper Paint Rock Lake
3. Lower Paint Rock Lake
Wilderness
Black Tooth Mtn
13005
Cloud Peak
Highland Park
Winnie Lake
Kearney Lake
Beaver Lake
Penrose Peak
12460
Frying Pan Lake
Gem Lake
Elk Lake
Willow Park Reservoir
8616
BIGHORN
NATIONAL
Roadless Balm of Gilead Area
Stone Mtn
7438
Bud Love Wildlife Habitat Management
Saddlestring
el 5480
HF Bar Ranch
UM Ranch
Flying E Ranch
Shell Creek School
Kearny
el 4657
Pilot Knob
4757
Fort Phil Kearny State Historic Site
Wagon Box Battle Monument (1867)
Fetterman Massacre Monument (1866)
Massacre Hill
4829
Sullivant Hill
5177
Lodge Trail Ridge
Story
el 5079
Story Fish Hatchery
KGK Tower
Rafter Y Ranch
Grandmas Mtn
6957
MONCREIFFE RIDGE
Little Goose Peak
9358
Banner
el 4617
Silver Lake
Big Horn
el 4081
Big Horn Events Center
Bird Farm
Bradford Brinton Memorial Museum
Quarter Circle A
Curlew Hill
4734
Snowmobile Parking
Bosin Rock
7707
Swamp Creek Hill
7669
Lamburger Rock
7881
Little Goose
Steep Grade
BEAVER CREEK HILLS
Beckton
el 4120
Kendrick Golf Club
Sheridan County Airport
4021
Mt Hope Cemetery
Kendrick Golf Course
Sheridan CC
Trail End SHS
Gillispie
Fort McKenzie VA Hospital
Thorne Rider Stadium
Sheridan County Museum
Welcome Center
Wyoming Game & Fish Office
Bighorn NF HQ
Bighorn Ranger District Office
Sheridan Junior College
Radio Facility
Hazel Peak
4218
Powder Horn Golf Club
Darlington
Paradise
Swaim Rd
Box Cross Rd
Hidden Hills
No Vehicles 11/16-5/14
Steerhead

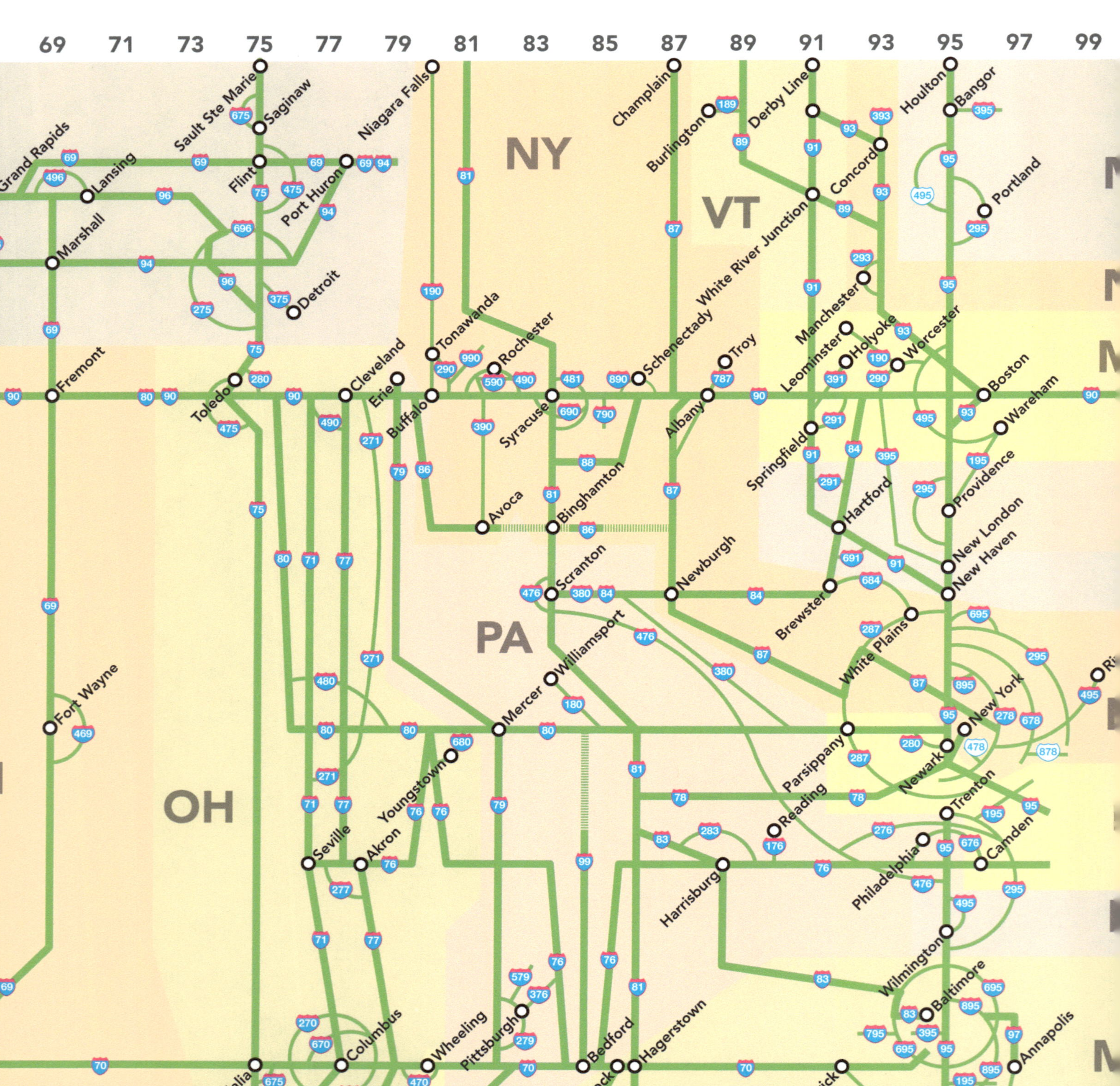

69
71
73
75
77
79
81
83
85
87
89
91
93
95
97
99
NY
VT
PA
OH
Sault Ste Marie
Saginaw
Niagara Falls
Champlain
Burlington
Derby Line
Houlton
Bangor
Grand Rapids
Lansing
Flint
Port Huron
Concord
Portland
Marshall
Detroit
White River Junction
Manchester
Tonawanda
Rochester
Schenectady
Troy
Leominster
Holyoke
Worcester
Boston
Fremont
Toledo
Cleveland
Erie
Buffalo
Syracuse
Albany
Springfield
Wareham
Providence
Hartford
Avoca
Binghamton
New London
New Haven
Scranton
Newburgh
Brewster
White Plains
Williamsport
Fort Wayne
Mercer
New York
Youngstown
Parsippany
Newark
Reading
Trenton
Seville
Akron
Harrisburg
Philadelphia
Camden
Wilmington
Baltimore
Columbus
Wheeling
Pittsburgh
Bedford
Hagerstown
Annapolis

Nat Case, Hedberg Maps, Inc.

A Numeric Topology of the United States Eisenhower Interstate Highway System

Software used: Adobe Illustrator

More information: www.hedbergmaps.com

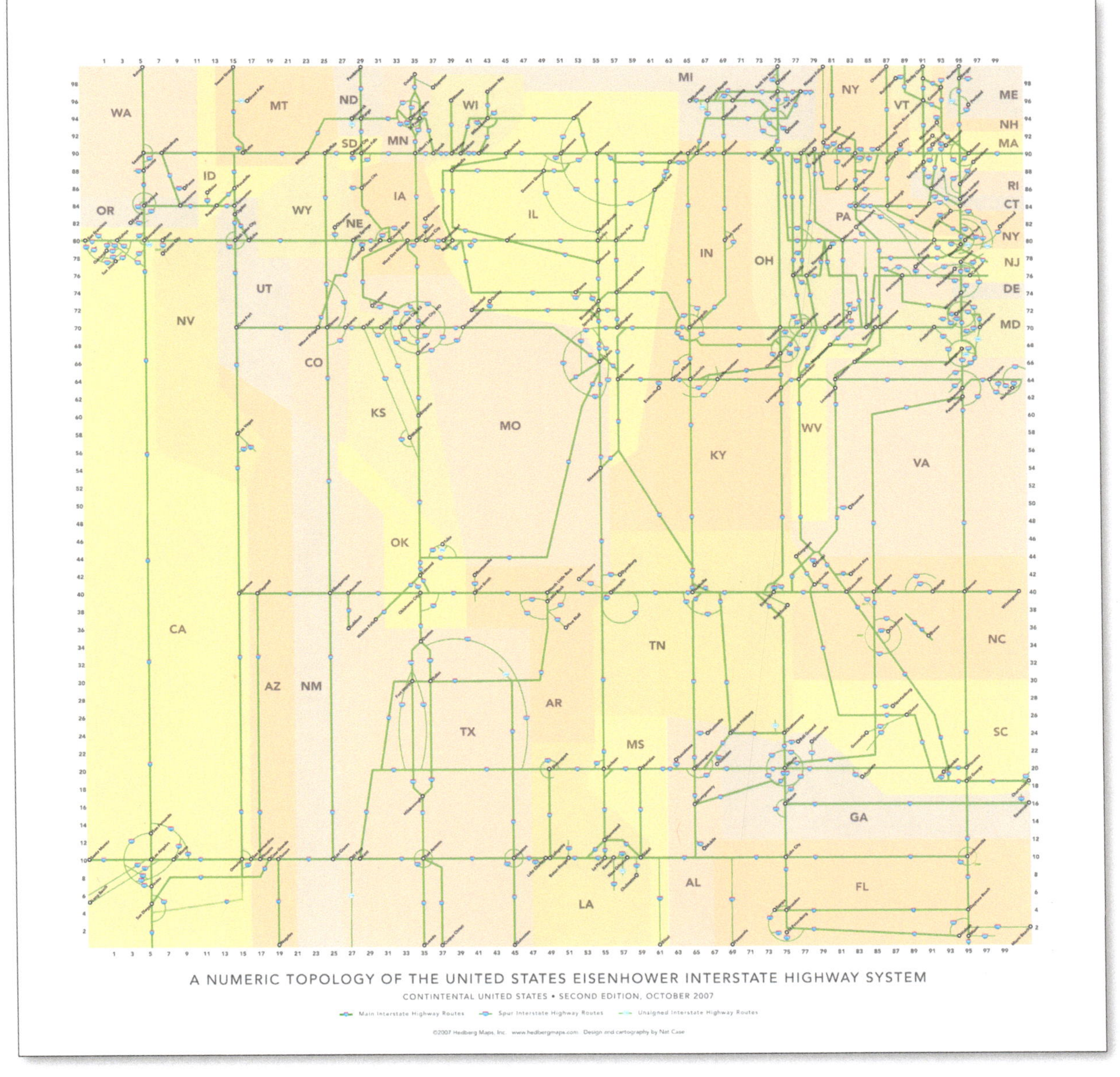

Jeff Clark, Clark Geomatics Corp.

Bivouac Backcountry Series: SW Garibaldi Park, BC, Canada

Software used: ESRI ArcGIS, Global Mapper, Adobe Photoshop, Adobe Illustrator, Adobe InDesign

Data sources: Landsat 7 ETM+, Canadian Digital Elevation Data, GISI Road Centreline Network, National Topographic Data Base

More information: www.clarkgeomatics.ca

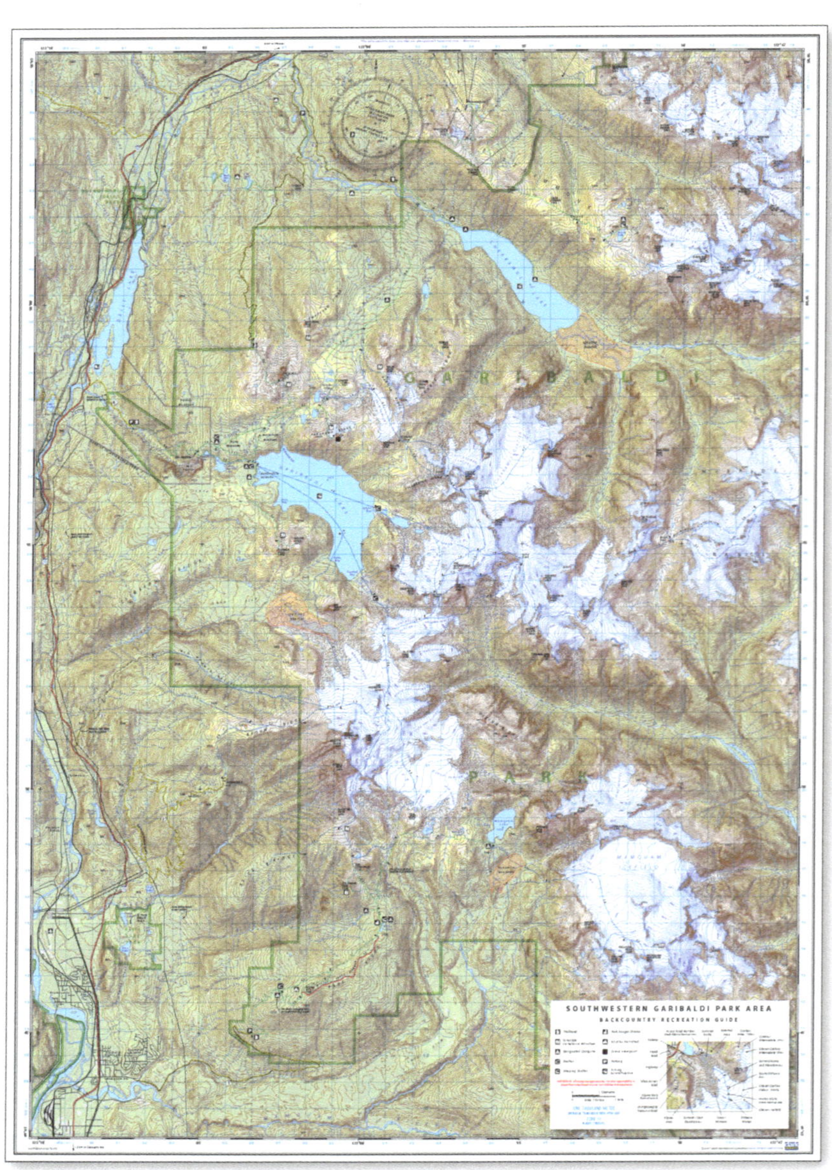

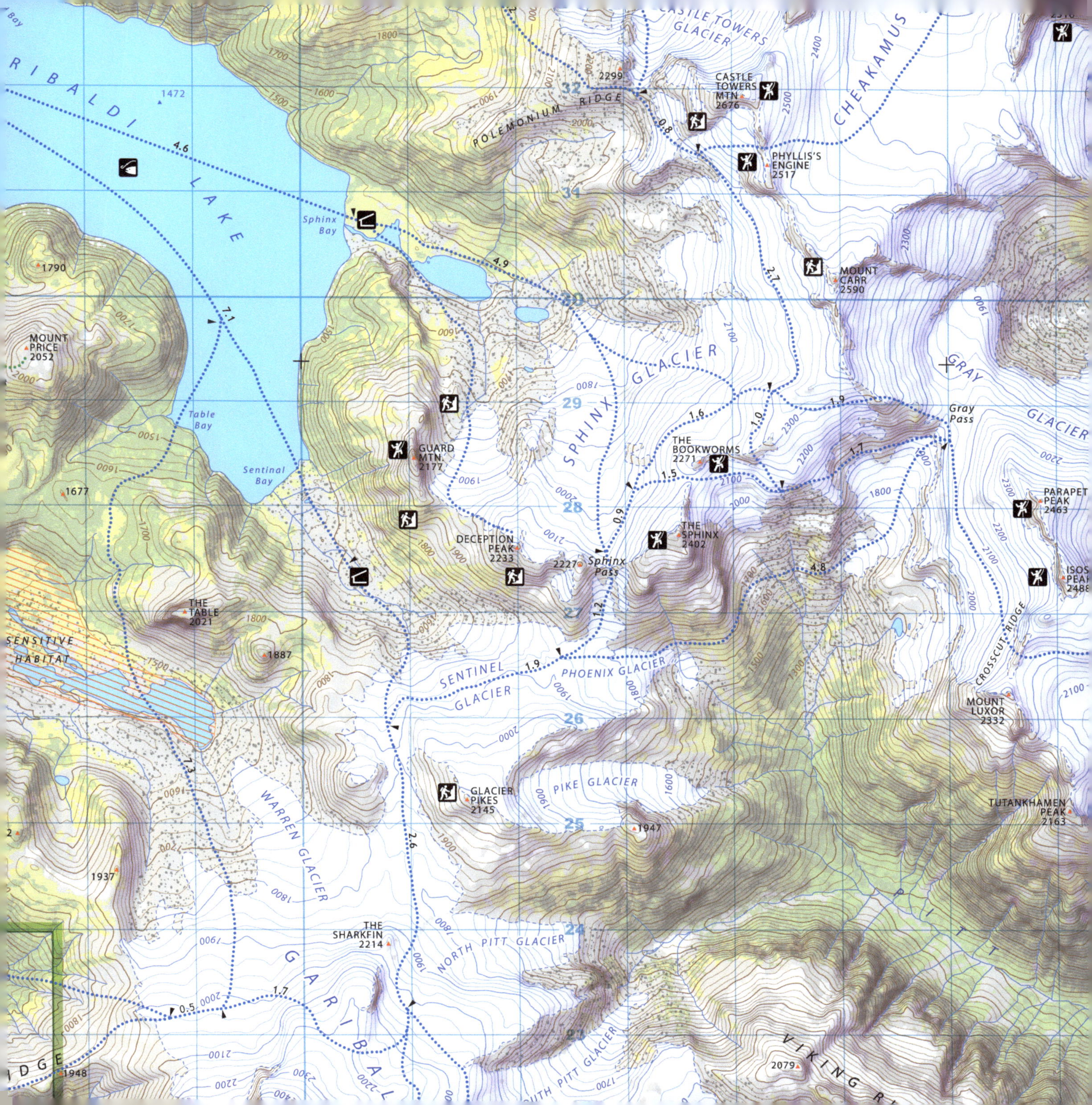

GARIBALDI LAKE
Sphinx Bay
Table Bay
Sentinal Bay
MOUNT PRICE 2052
1790
1677
1472
THE TABLE 2021
1887
SENSITIVE HABITAT
POLEMONIUM RIDGE
2299
CASTLE TOWERS GLACIER
CASTLE TOWERS MTN 2676
PHYLLIS'S ENGINE 2517
CHEAKAMUS
MOUNT CARR 2590
GUARD MTN 2177
DECEPTION PEAK 2233
2227
Sphinx Pass
SPHINX GLACIER
THE BOOKWORMS 2271
THE SPHINX 2402
GRAY GLACIER
Gray Pass
PARAPET PEAK 2463
SENTINEL GLACIER
PHOENIX GLACIER
CROSSCUT RIDGE
MOUNT LUXOR 2332
PIKE GLACIER
GLACIER PIKES 2145
1947
TUTANKHAMEN PEAK 2163
WARREN GLACIER
1937
THE SHARKFIN 2214
NORTH PITT GLACIER
PITT
VIKING
2079
1948

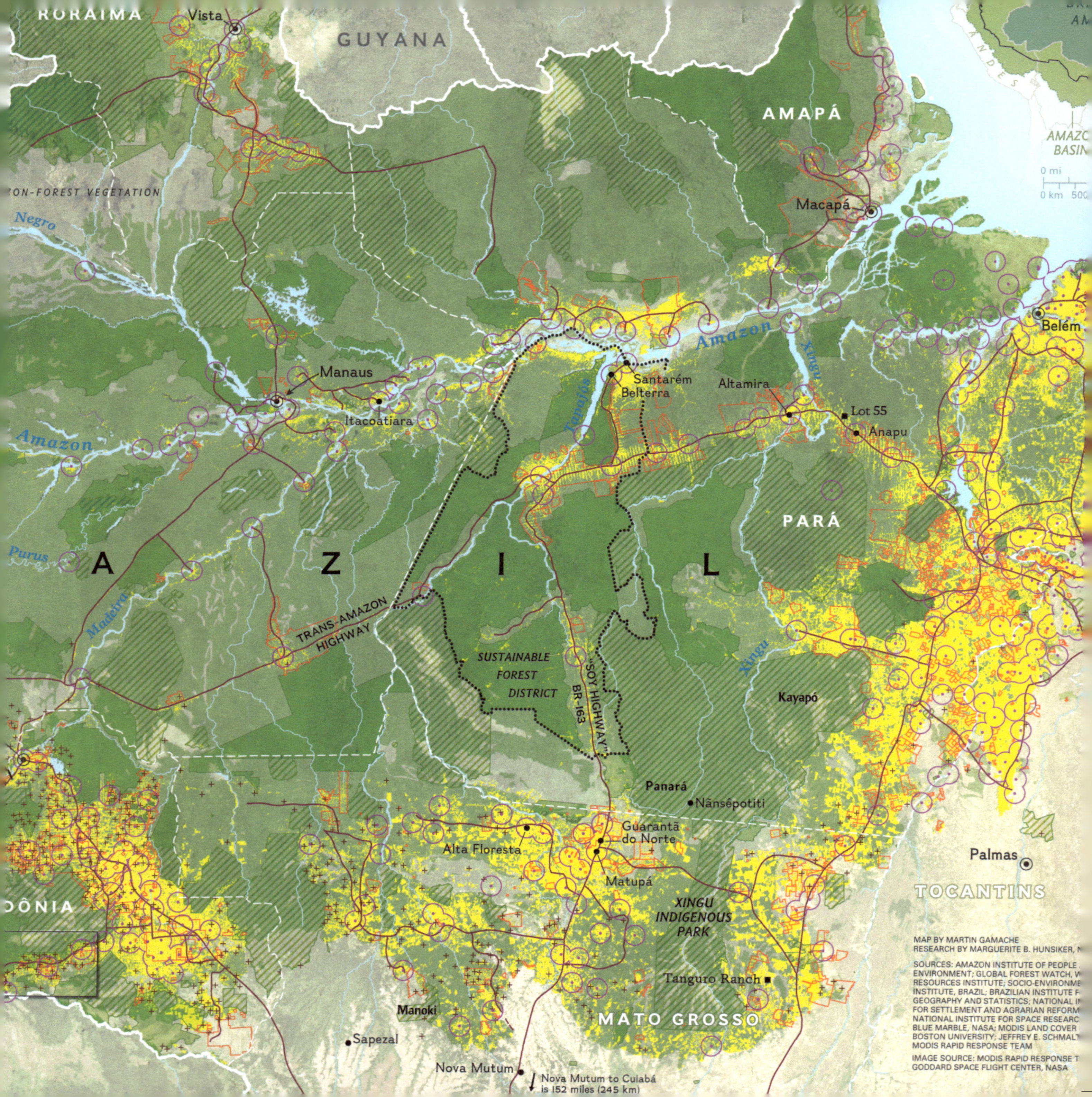
RORAIMA
Vista
GUYANA
AMAPÁ
Macapá
NON-FOREST VEGETATION
Negro
Amazon
Manaus
Itacoatiara
Santarém
Belterra
Tapajós
Altamira
Xingu
Lot 55
Anapu
Belém
PARÁ
Purus
Madeira
B R A Z I L
TRANS-AMAZON HIGHWAY
SUSTAINABLE FOREST DISTRICT
"SOY HIGHWAY" BR-163
Kayapó
Panará
Nânsêpotiti
Guarantã do Norte
Alta Floresta
Matupá
Palmas
TOCANTINS
XINGU INDIGENOUS PARK
Tanguro Ranch
MATO GROSSO
Manoki
Sapezal
Nova Mutum
Nova Mutum to Cuiabá is 152 miles (245 km)
RONDÔNIA
0 mi
0 km
MAP BY MARTIN GAMACHE
RESEARCH BY MARGUERITE B. HUNSIKER
SOURCES: AMAZON INSTITUTE OF PEOPLE
ENVIRONMENT; GLOBAL FOREST WATCH,
RESOURCES INSTITUTE; SOCIO-ENVIRONME
INSTITUTE, BRAZIL; BRAZILIAN INSTITUTE
GEOGRAPHY AND STATISTICS; NATIONAL
FOR SETTLEMENT AND AGRARIAN REFORM
NATIONAL INSTITUTE FOR SPACE RESEARC
BLUE MARBLE, NASA; MODIS LAND COVER
BOSTON UNIVERSITY; JEFFREY E. SCHMAL
MODIS RAPID RESPONSE TEAM
IMAGE SOURCE: MODIS RAPID RESPONSE
GODDARD SPACE FLIGHT CENTER, NASA

Martin Gamache, James McClelland, William E. McNulty, National Geographic Magazine

The Embattled Amazon (January, 2007)

Researcher: Marguerite B. Hunsiker

Software used: Manifold GIS, ESRI ArcMap, PCI Geomatics Focus, Adobe Illustrator, Adobe Photoshop

Data sources: Amazon Institute of People and the Environment; Global Forest Watch, World Resources Institute; Socio-Environmental Institute, Brazil; Brazilian Institute for Geography and Statistics; National Institute for Settlement and Agrarian Reform, Brazil; National Institute for Space Research, Brazil; Blue Marble, NASA; MODIS Land Cover Product, Boston University; Jeffrey E. Schmaltz, MODIS Rapid Response Team

Image Source: MODIS Rapid Response Team, Goddard Space Flight Center, NASA

More information: ngm.nationalgeographic.com/ngm

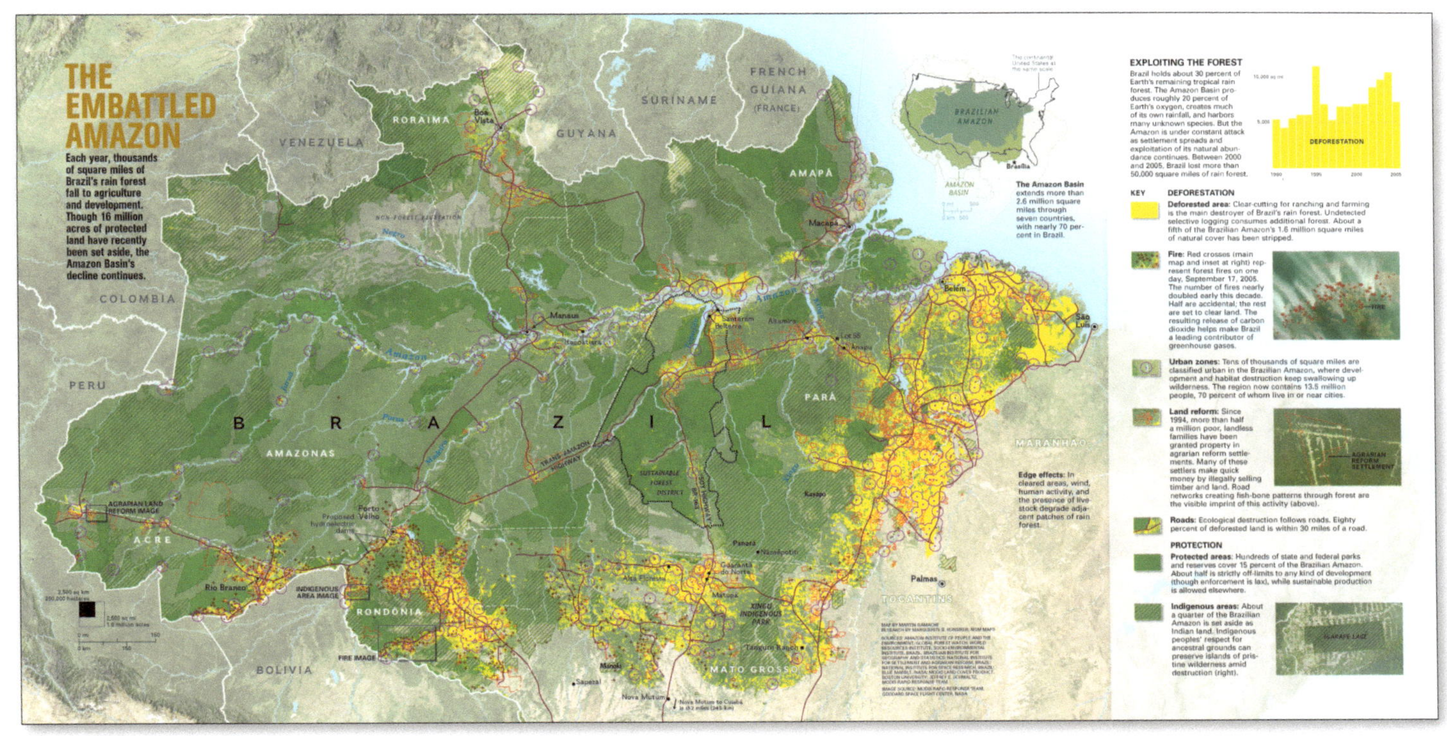

Francois Goulet, QA International

L'Amérique du Nord économique
("North America Economic Map")
From Mon atlas du monde ("My Atlas of the World")*
Cartography by Francois Goulet, Design by QA International Team

Software used: Avenza MAPublisher, Adobe Illustrator, ESRI ArcMap
More information: www.francoisgoulet.com

**Titre provisoire/Temporary title*

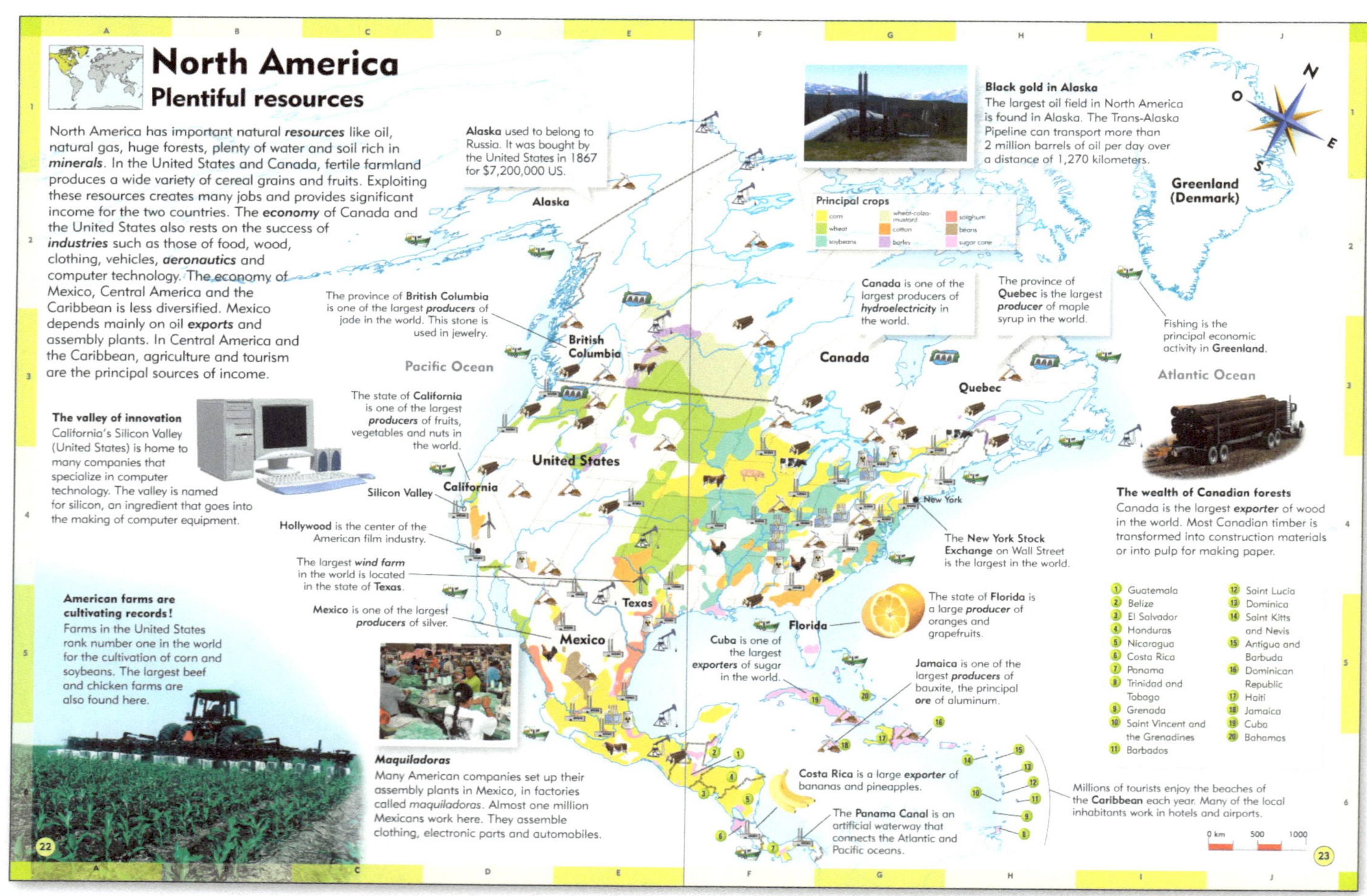

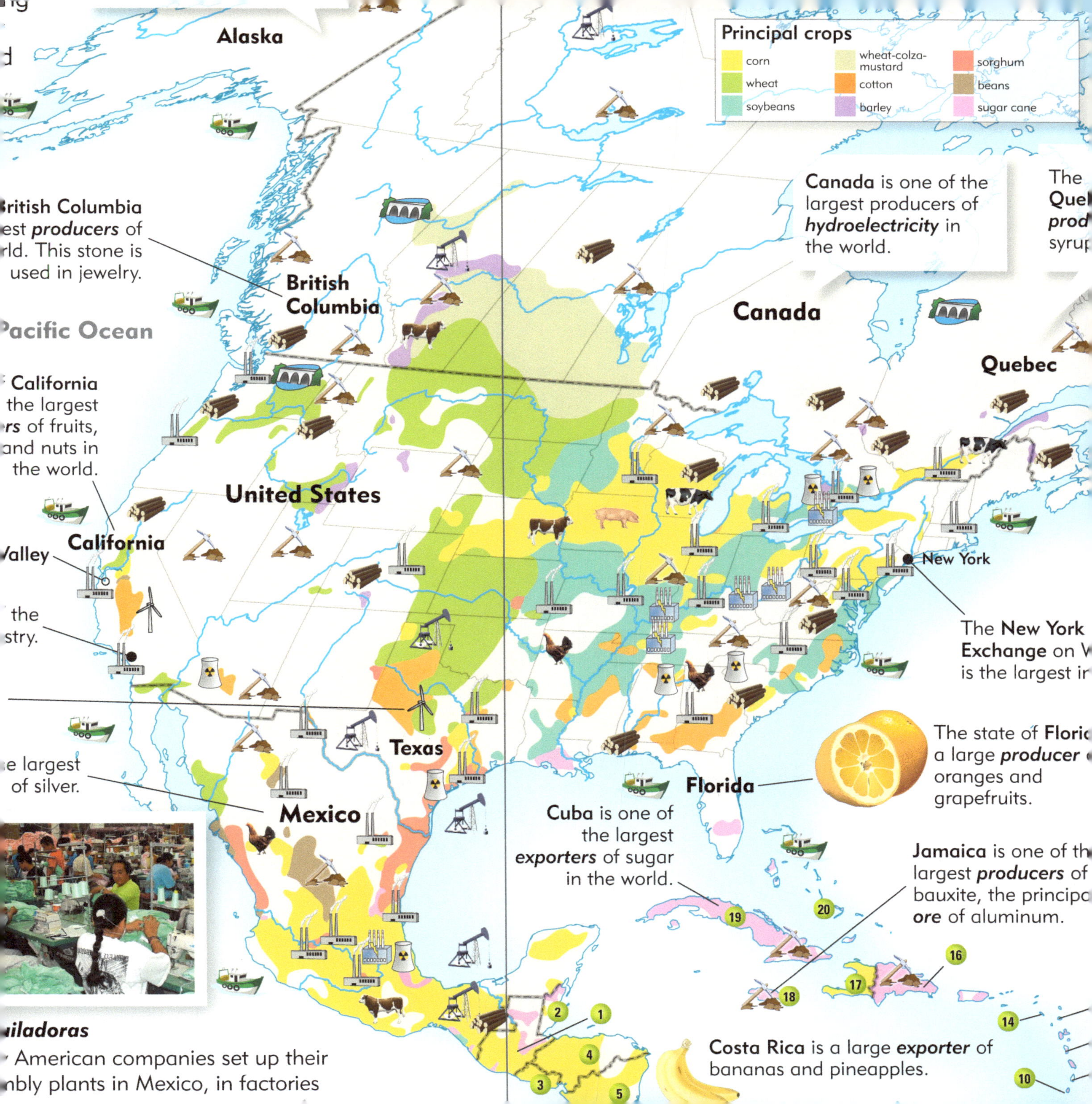

Alaska
Principal crops
corn
wheat
soybeans
wheat-colza-mustard
cotton
barley
sorghum
beans
sugar cane
ritish Columbia
est producers of
rld. This stone is
used in jewelry.
British Columbia
Pacific Ocean
Canada
Canada is one of the largest producers of hydroelectricity in the world.
The
Que
prod
syrup
Quebec
California
the largest
rs of fruits,
and nuts in
the world.
United States
California
alley
the
stry.
New York
The New York
Exchange on W
is the largest in
Texas
e largest
of silver.
Mexico
Florida
The state of Florid
a large producer
oranges and
grapefruits.
Cuba is one of the largest exporters of sugar in the world.
Jamaica is one of th
largest producers of
bauxite, the principa
ore of aluminum.
iiladoras
American companies set up their
nbly plants in Mexico, in factories
Costa Rica is a large exporter of bananas and pineapples.
19
20
16
17
18
14
10
1
2
3
4
5

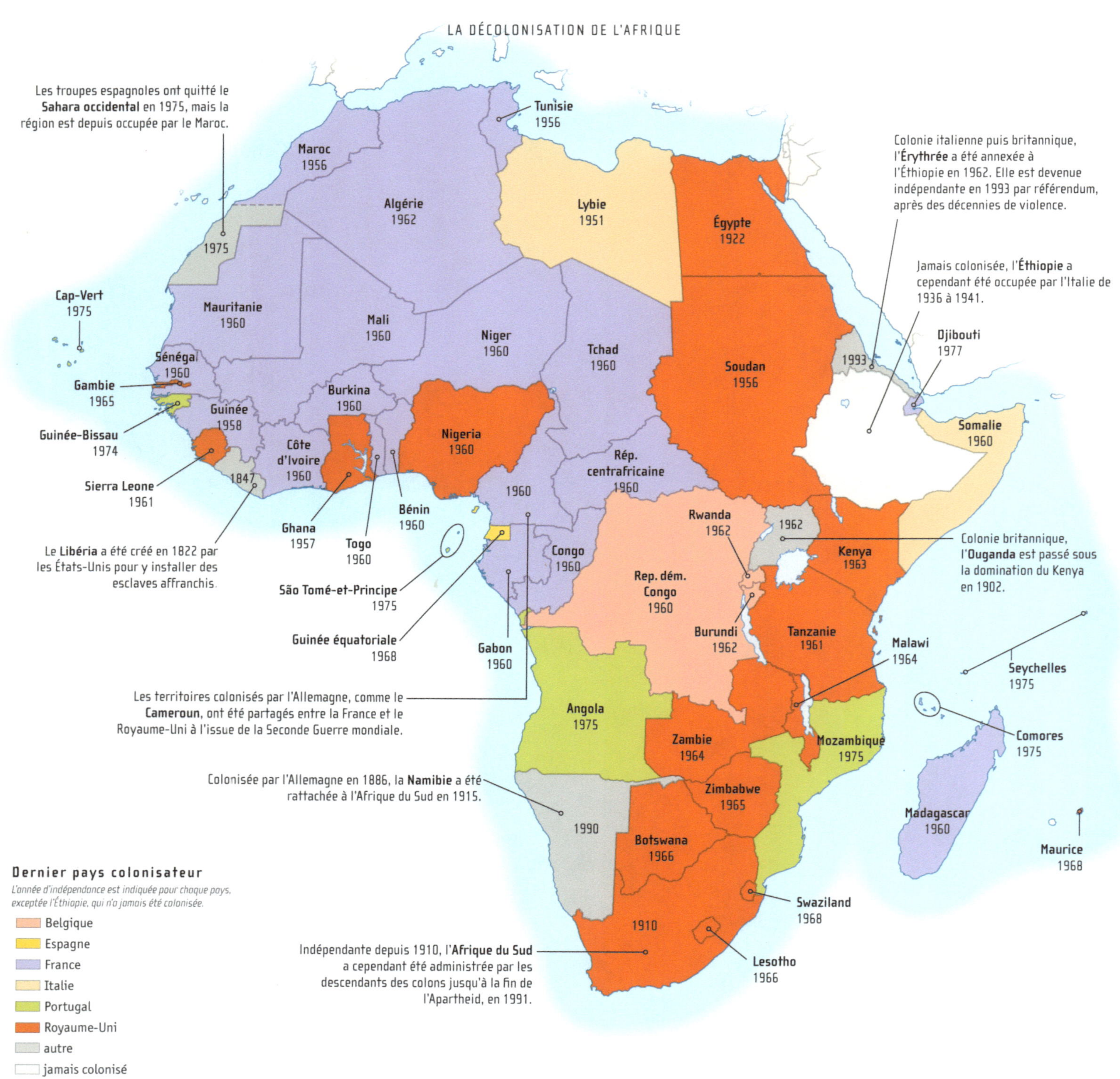

LA DÉCOLONISATION DE L'AFRIQUE
Les troupes espagnoles ont quitté le **Sahara occidental** en 1975, mais la région est depuis occupée par le Maroc.
Tunisie 1956
Maroc 1956
Algérie 1962
Lybie 1951
Égypte 1922
1975
Colonie italienne puis britannique, l'**Érythrée** a été annexée à l'Éthiopie en 1962. Elle est devenue indépendante en 1993 par référendum, après des décennies de violence.
Jamais colonisée, l'**Éthiopie** a cependant été occupée par l'Italie de 1936 à 1941.
Cap-Vert 1975
Mauritanie 1960
Mali 1960
Niger 1960
Tchad 1960
Soudan 1956
1993
Djibouti 1977
Sénégal 1960
Gambie 1965
Guinée 1958
Burkina 1960
Nigeria 1960
Somalie 1960
Guinée-Bissau 1974
Sierra Leone 1961
1847
Côte d'Ivoire 1960
Rép. centrafricaine 1960
1960
Ghana 1957
Togo 1960
Bénin 1960
Rwanda 1962
1962
Kenya 1963
Colonie britannique, l'**Ouganda** est passé sous la domination du Kenya en 1902.
Le **Libéria** a été créé en 1822 par les États-Unis pour y installer des esclaves affranchis.
São Tomé-et-Principe 1975
Congo 1960
Rep. dém. Congo 1960
Guinée équatoriale 1968
Gabon 1960
Burundi 1962
Tanzanie 1961
Malawi 1964
Seychelles 1975
Les territoires colonisés par l'Allemagne, comme le **Cameroun**, ont été partagés entre la France et le Royaume-Uni à l'issue de la Seconde Guerre mondiale.
Angola 1975
Zambie 1964
Mozambique 1975
Comores 1975
Colonisée par l'Allemagne en 1886, la **Namibie** a été rattachée à l'Afrique du Sud en 1915.
Zimbabwe 1965
1990
Botswana 1966
Madagascar 1960
Maurice 1968
Swaziland 1968
1910
Lesotho 1966
Indépendante depuis 1910, l'**Afrique du Sud** a cependant été administrée par les descendants des colons jusqu'à la fin de l'Apartheid, en 1991.
Dernier pays colonisateur
L'année d'indépendance est indiquée pour chaque pays, exceptée l'Éthiopie, qui n'a jamais été colonisée.
Belgique
Espagne
France
Italie
Portugal
Royaume-Uni
autre
jamais colonisé

Francois Goulet, QA International

L'Afrique ("Africa")

From Atlas visuel du monde ("Visual World Atlas")

Cartography by Francois Goulet, Design by QA International Team

Software used: Avenza MAPublisher, Adobe Illustrator, ESRI ArcMap

More information: www.francoisgoulet.com

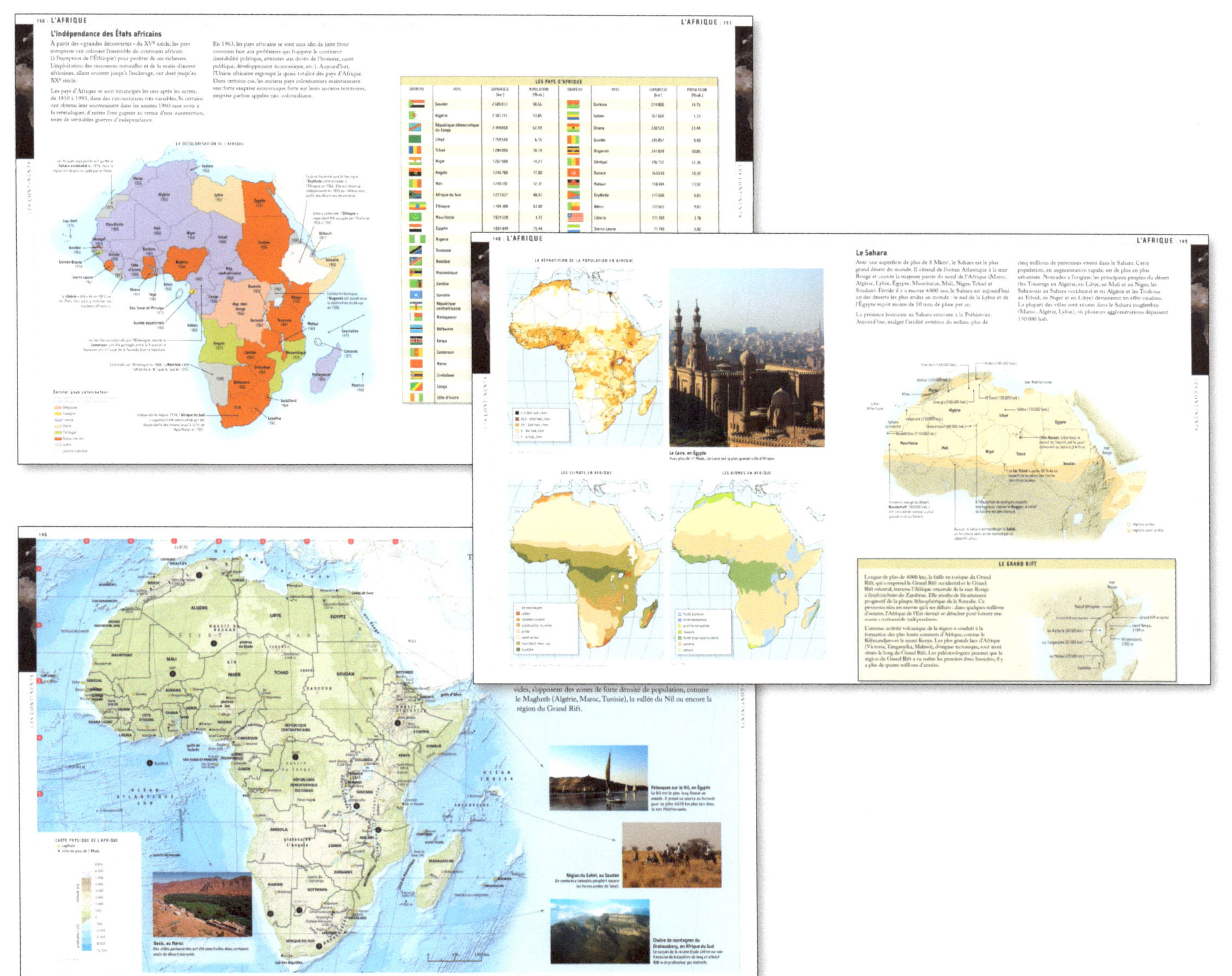

Casey Greene

Big Island Pop-Top Amenity Guide

Software used: Avenza MAPublisher, Adobe Illustrator, Adobe Photoshop, ESRI ArcGIS Desktop

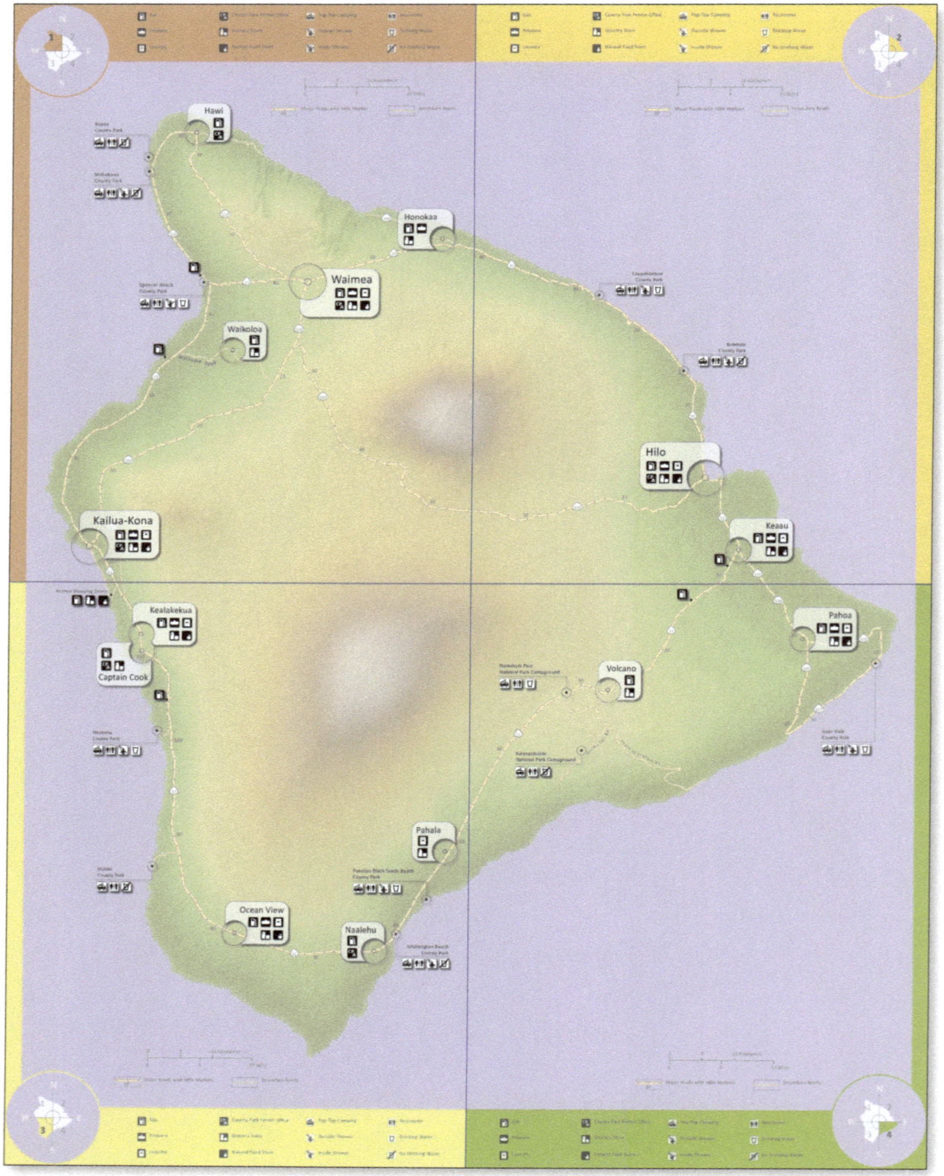

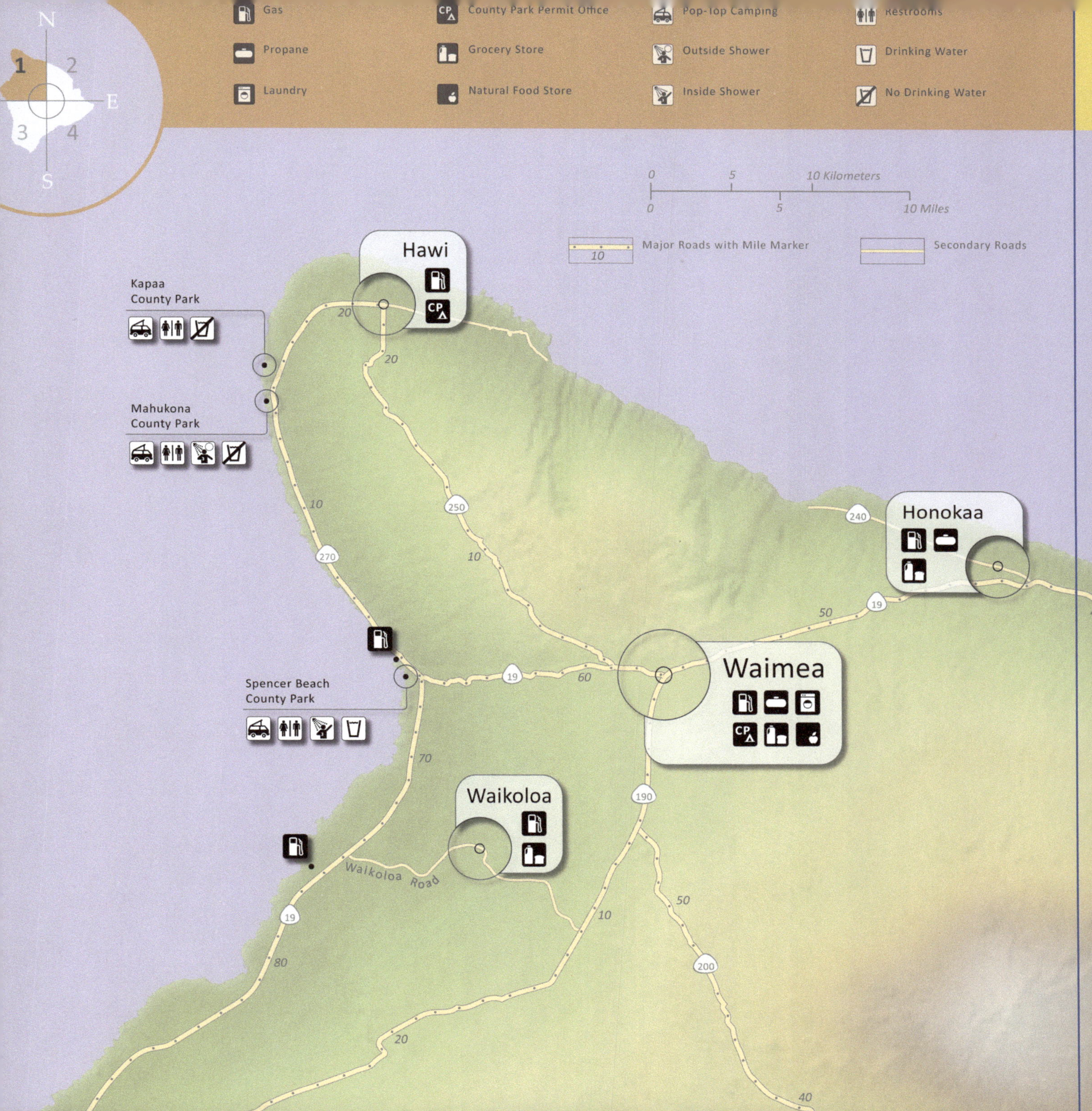
N
E
S
1
2
3
4
Gas
County Park Permit Office
Pop-Top Camping
Restrooms
Propane
Grocery Store
Outside Shower
Drinking Water
Laundry
Natural Food Store
Inside Shower
No Drinking Water
0 5 10 Kilometers
0 5 10 Miles
Major Roads with Mile Marker
Secondary Roads
Hawi
Kapaa County Park
Mahukona County Park
Spencer Beach County Park
Honokaa
Waimea
Waikoloa
Waikoloa Road
270
250
240
19
190
200

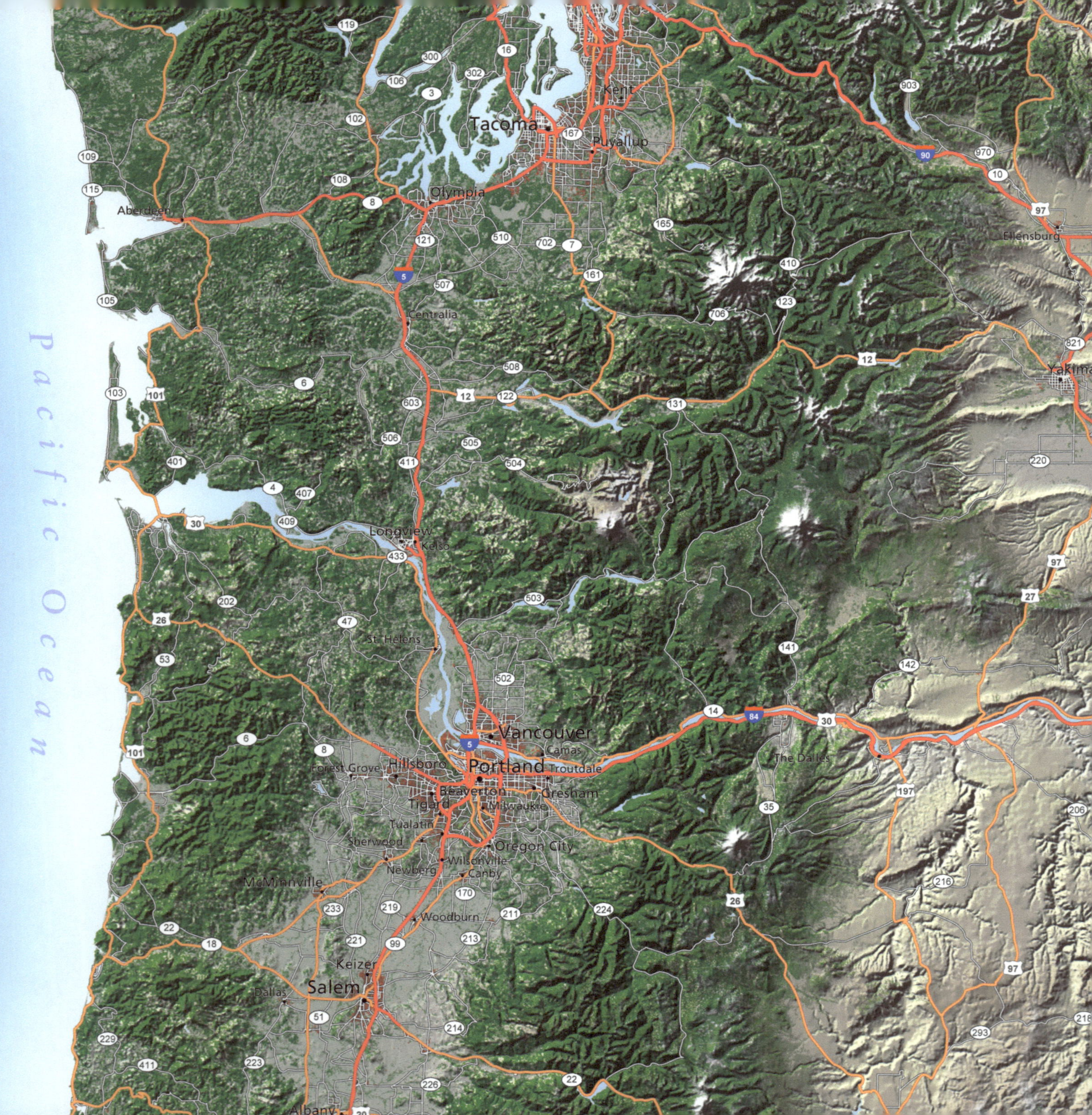

Pacific Ocean
Tacoma
Kent
Puyallup
Olympia
Aberdeen
Centralia
Ellensburg
Yakima
Longview
Kelso
St. Helens
Vancouver
Camas
Portland
Troutdale
Gresham
Hillsboro
Forest Grove
Beaverton
Tigard
Milwaukie
Tualatin
Sherwood
Oregon City
Wilsonville
Newberg
Canby
McMinnville
Woodburn
Keizer
Salem
Dallas
Albany
The Dalles

Matthew Hampton, Oregon Metro

Pacific Northwest

Software used: ESRI ArcGIS, Adobe Illustrator, Adobe Photoshop

Data sources: Metro's Regional Land Information System (R.L.I.S.), National Land Cover Dataset, ODOT, ESRI.

More information: www.oregonmetro.gov

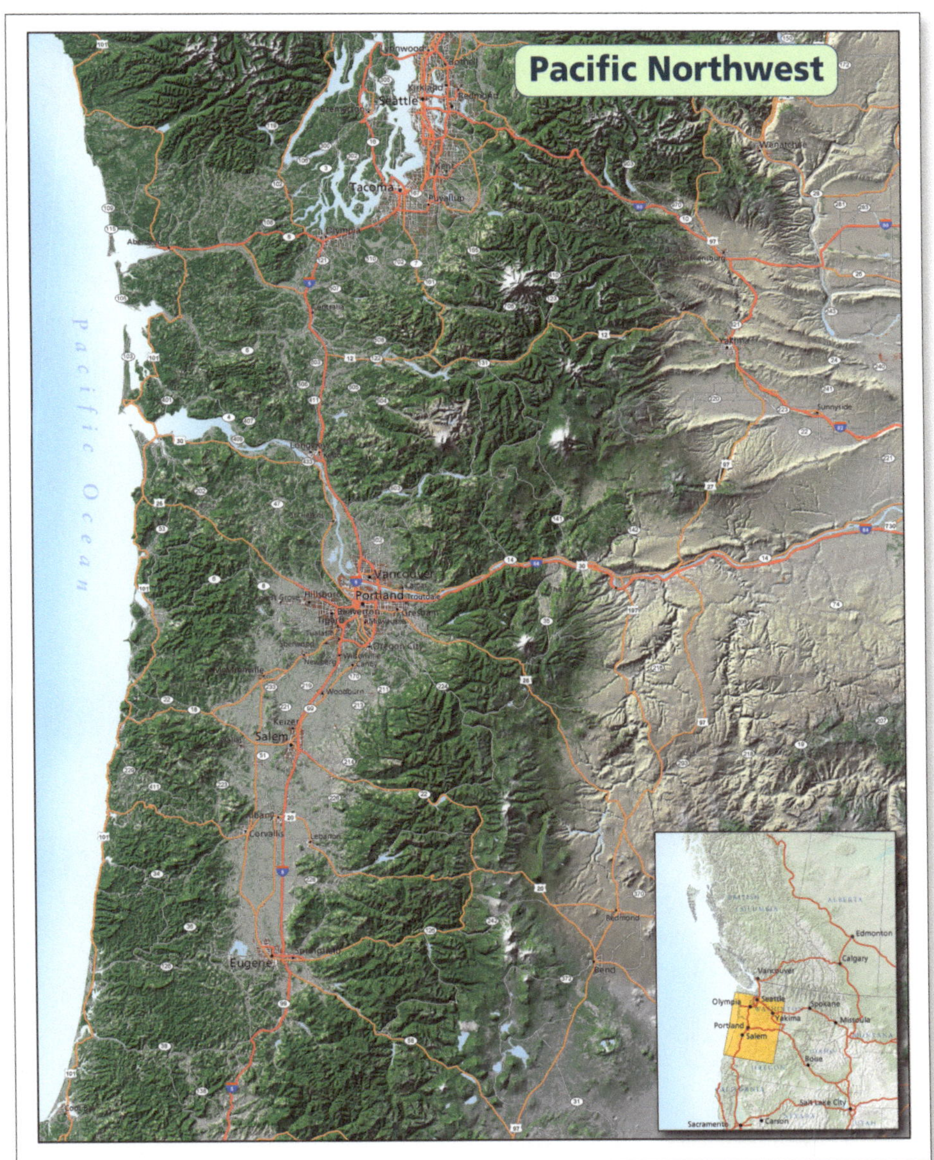

Danielle Hartman, Halcrow, Inc.

GVA Williams Manhattan

Software used: ESRI ArcGIS with Maplex, Adobe Illustrator

Data sources: Customized from public and private sources

More information: www.halcrow.com/maps

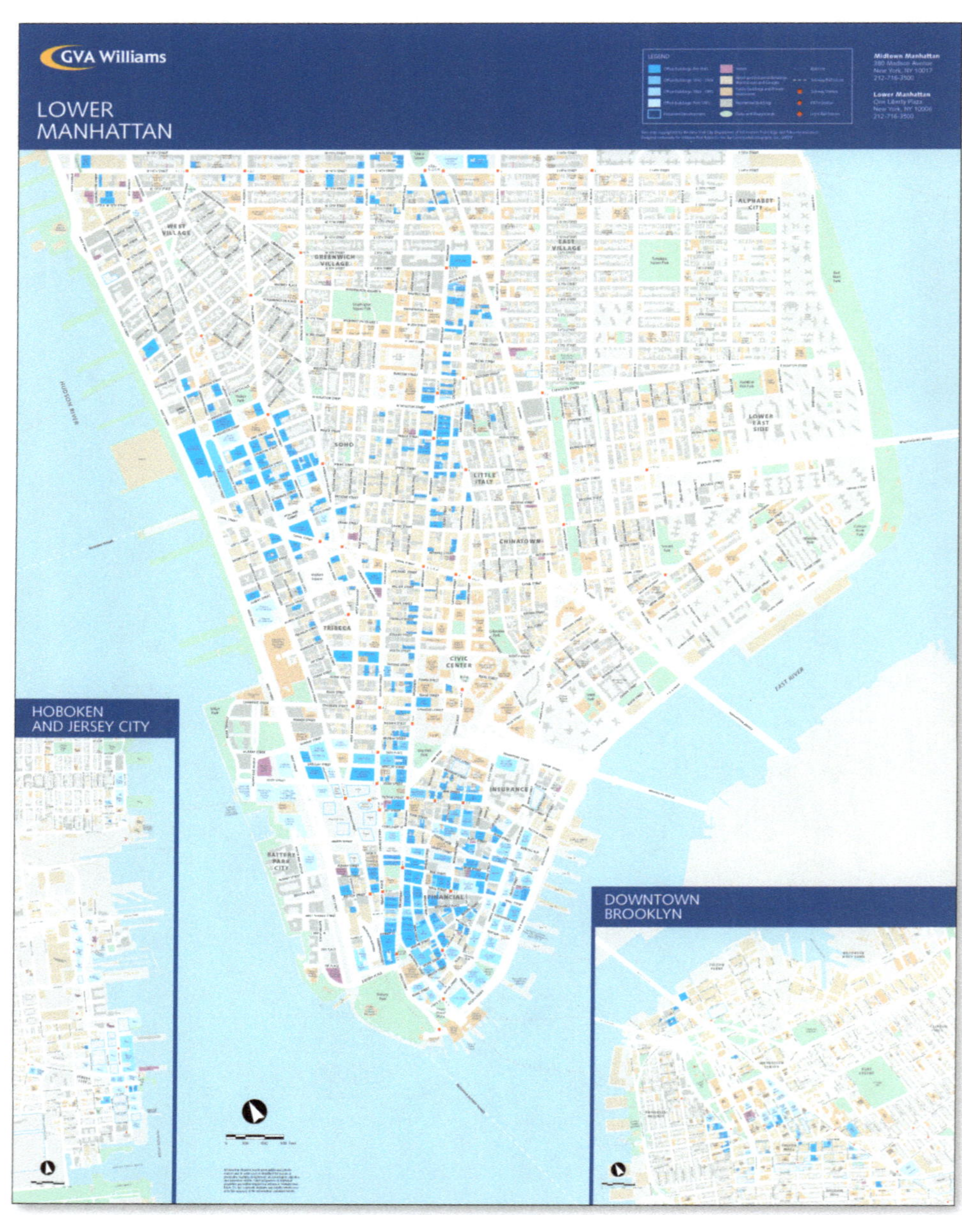

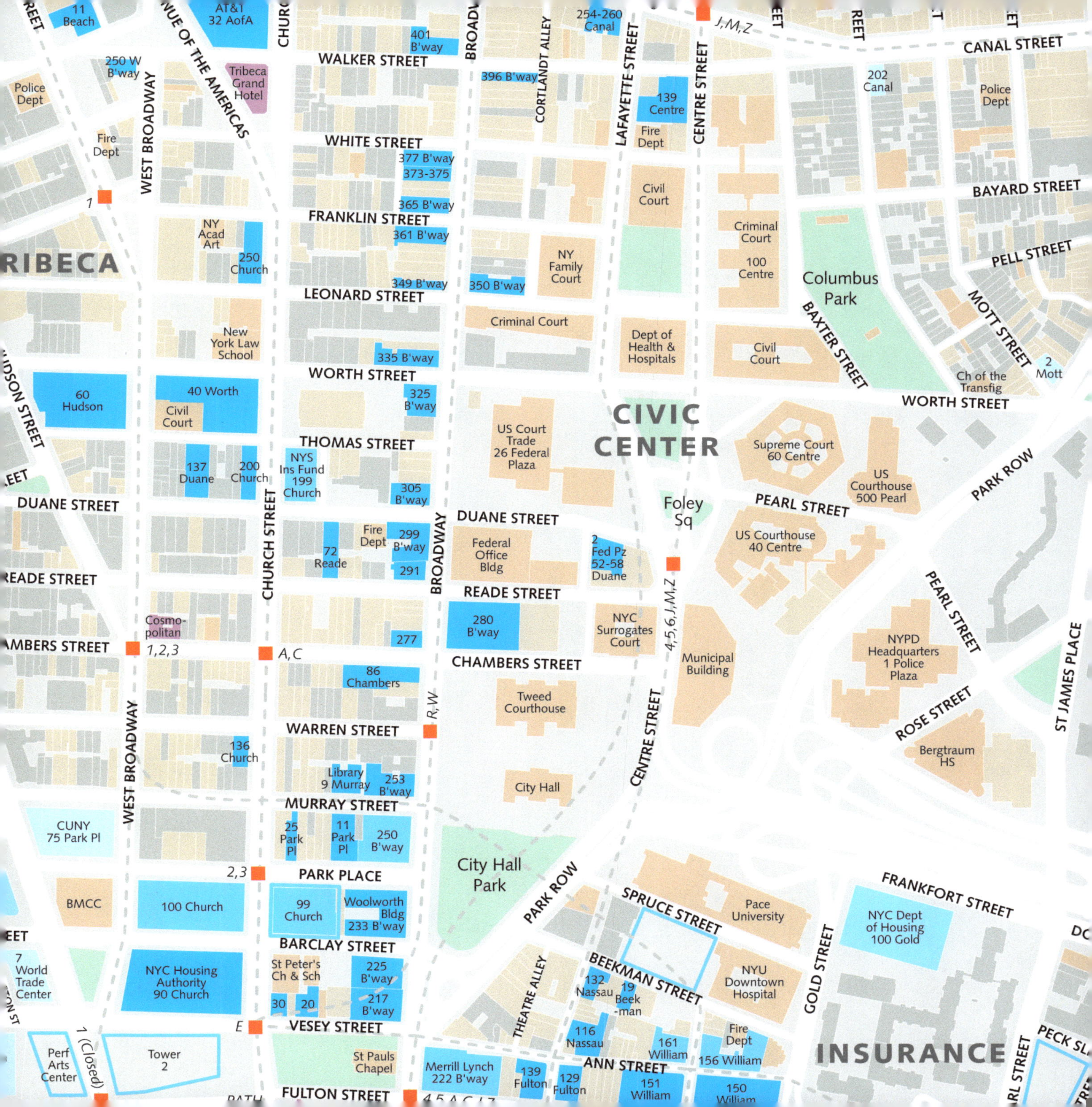

TRIBECA
CIVIC CENTER
INSURANCE
CANAL STREET
WALKER STREET
WHITE STREET
FRANKLIN STREET
LEONARD STREET
WORTH STREET
THOMAS STREET
DUANE STREET
READE STREET
CHAMBERS STREET
WARREN STREET
MURRAY STREET
PARK PLACE
BARCLAY STREET
VESEY STREET
FULTON STREET
BAYARD STREET
PELL STREET
MOTT STREET
BAXTER STREET
PEARL STREET
PARK ROW
ROSE STREET
ST JAMES PLACE
SPRUCE STREET
BEEKMAN STREET
FRANKFORT STREET
GOLD STREET
ANN STREET
THEATRE ALLEY
CENTRE STREET
LAFAYETTE STREET
CORTLANDT ALLEY
BROADWAY
CHURCH STREET
WEST BROADWAY
AVENUE OF THE AMERICAS
HUDSON STREET
PECK SL
11 Beach
AT&T 32 AofA
401 B'way
254-260 Canal
J,M,Z
250 W B'way
Tribeca Grand Hotel
396 B'way
139 Centre
202 Canal
Police Dept
Fire Dept
377 B'way
373-375
365 B'way
361 B'way
349 B'way
350 B'way
Civil Court
Criminal Court
100 Centre
NY Family Court
NY Acad Art
250 Church
Columbus Park
New York Law School
335 B'way
Dept of Health & Hospitals
Ch of the Transfig
2 Mott
60 Hudson
40 Worth
325 B'way
US Court Trade 26 Federal Plaza
Supreme Court 60 Centre
US Courthouse 500 Pearl
137 Duane
200 Church
NYS Ins Fund 199 Church
305 B'way
Foley Sq
US Courthouse 40 Centre
72 Reade
299 B'way
291
Federal Office Bldg
2 Fed Pz 52-58 Duane
Cosmo-politan
1,2,3
A,C
277
280 B'way
NYC Surrogates Court
4,5,6,J,M,Z
Municipal Building
NYPD Headquarters 1 Police Plaza
86 Chambers
Tweed Courthouse
136 Church
R,W
Library 9 Murray
253 B'way
City Hall
Bergtraum HS
CUNY 75 Park Pl
25 Park Pl
11 Park Pl
250 B'way
City Hall Park
2,3
BMCC
100 Church
99 Church
Woolworth Bldg
233 B'way
Pace University
NYC Dept of Housing 100 Gold
7 World Trade Center
NYC Housing Authority 90 Church
St Peter's Ch & Sch
225 B'way
217 B'way
30
20
132 Nassau
19 Beek -man
NYU Downtown Hospital
E
116 Nassau
161 William
156 William
Perf Arts Center
1 (Closed)
Tower 2
St Pauls Chapel
Merrill Lynch 222 B'way
139 Fulton
129 Fulton
151 William
150 William

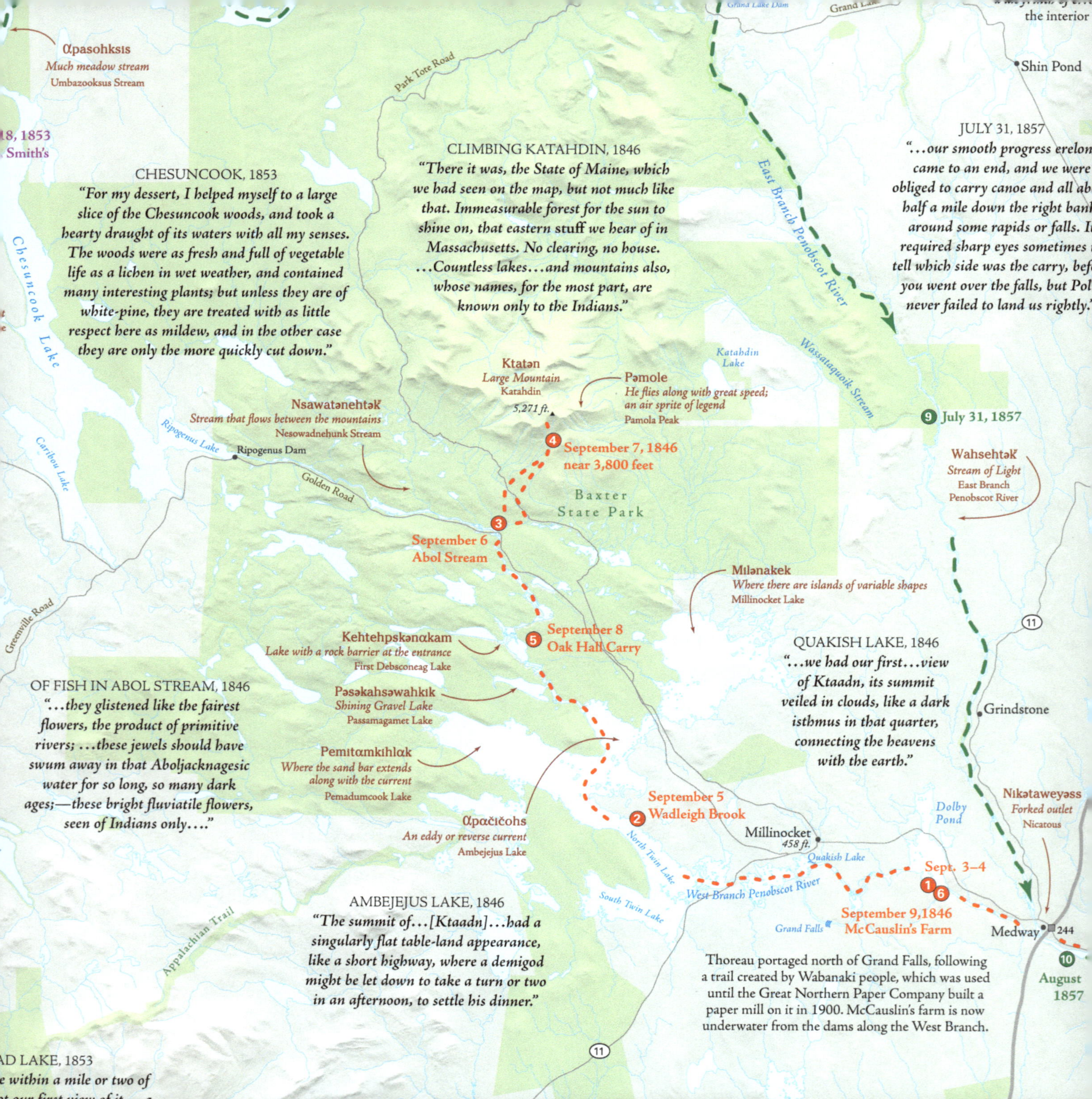

ɑpasohksis
Much meadow stream
Umbazooksus Stream
18, 1853
Smith's
CHESUNCOOK, 1853
"For my dessert, I helped myself to a large slice of the Chesuncook woods, and took a hearty draught of its waters with all my senses. The woods were as fresh and full of vegetable life as a lichen in wet weather, and contained many interesting plants; but unless they are of white-pine, they are treated with as little respect here as mildew, and in the other case they are only the more quickly cut down."
Chesuncook Lake
Park Tote Road
CLIMBING KATAHDIN, 1846
"There it was, the State of Maine, which we had seen on the map, but not much like that. Immeasurable forest for the sun to shine on, that eastern stuff we hear of in Massachusetts. No clearing, no house. ...Countless lakes...and mountains also, whose names, for the most part, are known only to the Indians."
Grand Lake Dam
Grand Lake
the interior
Shin Pond
JULY 31, 1857
"...our smooth progress erelon came to an end, and we were obliged to carry canoe and all ab half a mile down the right ban around some rapids or falls. I required sharp eyes sometimes tell which side was the carry, bef you went over the falls, but Pol never failed to land us rightly."
East Branch Penobscot River
Wassataquoik Stream
Katahdin Lake
Ktatən
Large Mountain
Katahdin
5,271 ft.
Pəmole
He flies along with great speed; an air sprite of legend
Pamola Peak
Nsawatənehtək
Stream that flows between the mountains
Nesowadnehunk Stream
Ripogenus Lake
Ripogenus Dam
Caribou Lake
Golden Road
4 September 7, 1846 near 3,800 feet
Baxter State Park
9 July 31, 1857
Wahsehtək
Stream of Light
East Branch Penobscot River
3 September 6 Abol Stream
Milənakek
Where there are islands of variable shapes
Millinocket Lake
Greenville Road
5 September 8 Oak Hall Carry
Kehtehpskənɑkam
Lake with a rock barrier at the entrance
First Debsconeag Lake
11
QUAKISH LAKE, 1846
"...we had our first...view of Ktaadn, its summit veiled in clouds, like a dark isthmus in that quarter, connecting the heavens with the earth."
OF FISH IN ABOL STREAM, 1846
"...they glistened like the fairest flowers, the product of primitive rivers; ...these jewels should have swum away in that Aboljacknagesic water for so long, so many dark ages;—these bright fluviatile flowers, seen of Indians only...."
Pəsəkahsəwahkɪk
Shining Gravel Lake
Passamagamet Lake
Grindstone
Pemɪtɑmkɪhlɑk
Where the sand bar extends along with the current
Pemadumcook Lake
2 September 5 Wadleigh Brook
Dolby Pond
Nikətaweyəss
Forked outlet
Nicatous
ɑpačičohs
An eddy or reverse current
Ambejejus Lake
Millinocket
458 ft.
Quakish Lake
North Twin Lake
Sept. 3–4
1 6
West Branch Penobscot River
South Twin Lake
September 9, 1846 McCauslin's Farm
Grand Falls
Medway
244
10 August 1857
Appalachian Trail
AMBEJEJUS LAKE, 1846
"The summit of...[Ktaadn]...had a singularly flat table-land appearance, like a short highway, where a demigod might be let down to take a turn or two in an afternoon, to settle his dinner."
Thoreau portaged north of Grand Falls, following a trail created by Wabanaki people, which was used until the Great Northern Paper Company built a paper mill on it in 1900. McCauslin's farm is now underwater from the dams along the West Branch.
11
AD LAKE, 1853
e within a mile or two of
ot our first view of it.—a

Michael Hermann, University of Maine Canadian-American Center

Thoreau-Wabanaki Trail

Cartography by Michael Hermann, senior cartographer, University of Maine Canadian-American Center. Co-designers James Eric Francis, Sr. and Margaret Pearce.

Software used: Adobe Illustrator, Avenza MAPublisher, Adobe Photoshop, Natural Scene Designer

More information: www.thoreauwabanakitrail.org

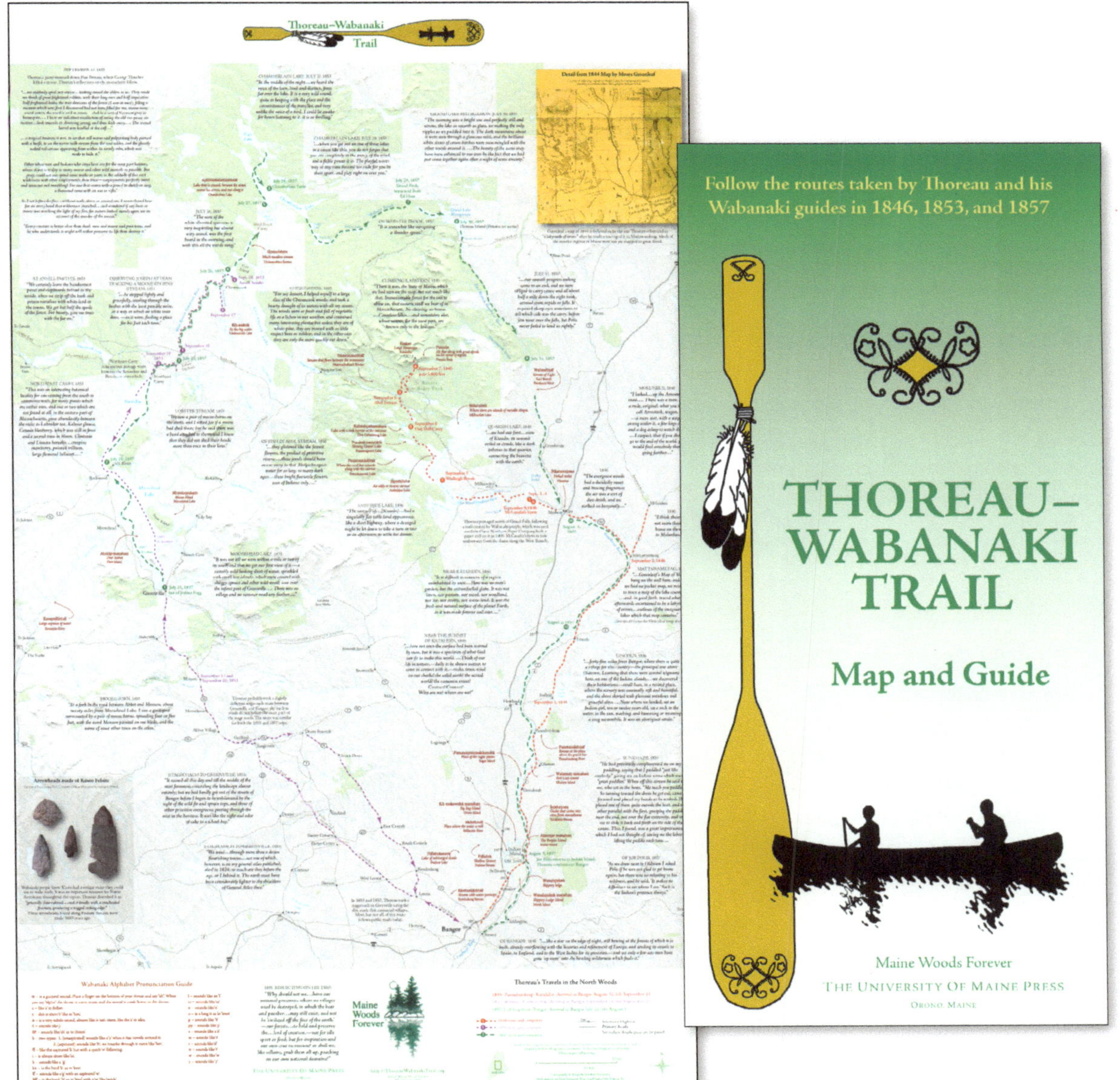

Rob James

Papahānaumokuākea Marine National Monument

Software used: ESRI ArcGIS, Adobe Illustrator, Adobe Photoshop

Data sources: NOAA National Marine Sanctuary Program

More information: hawaiireef.noaa.gov

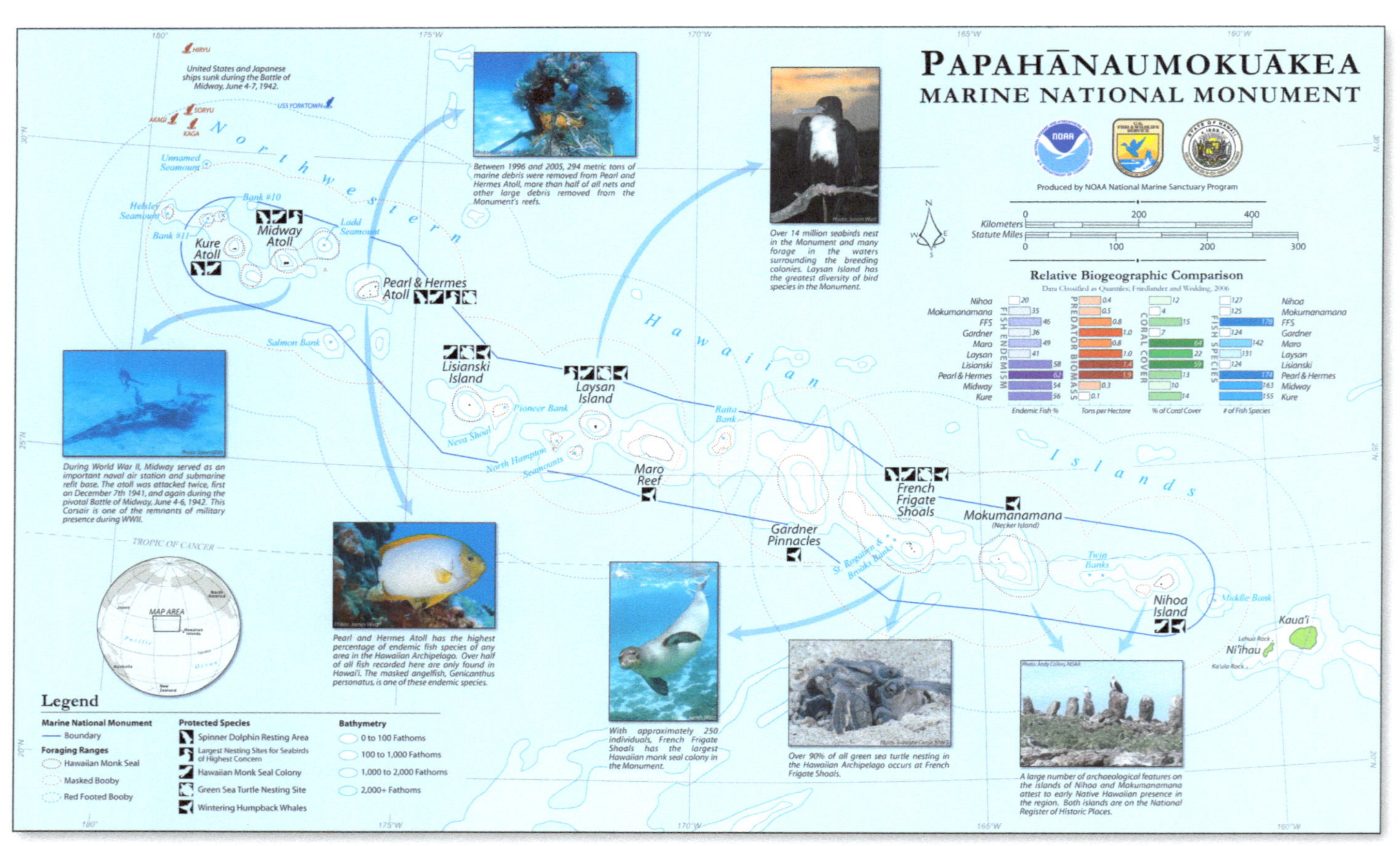

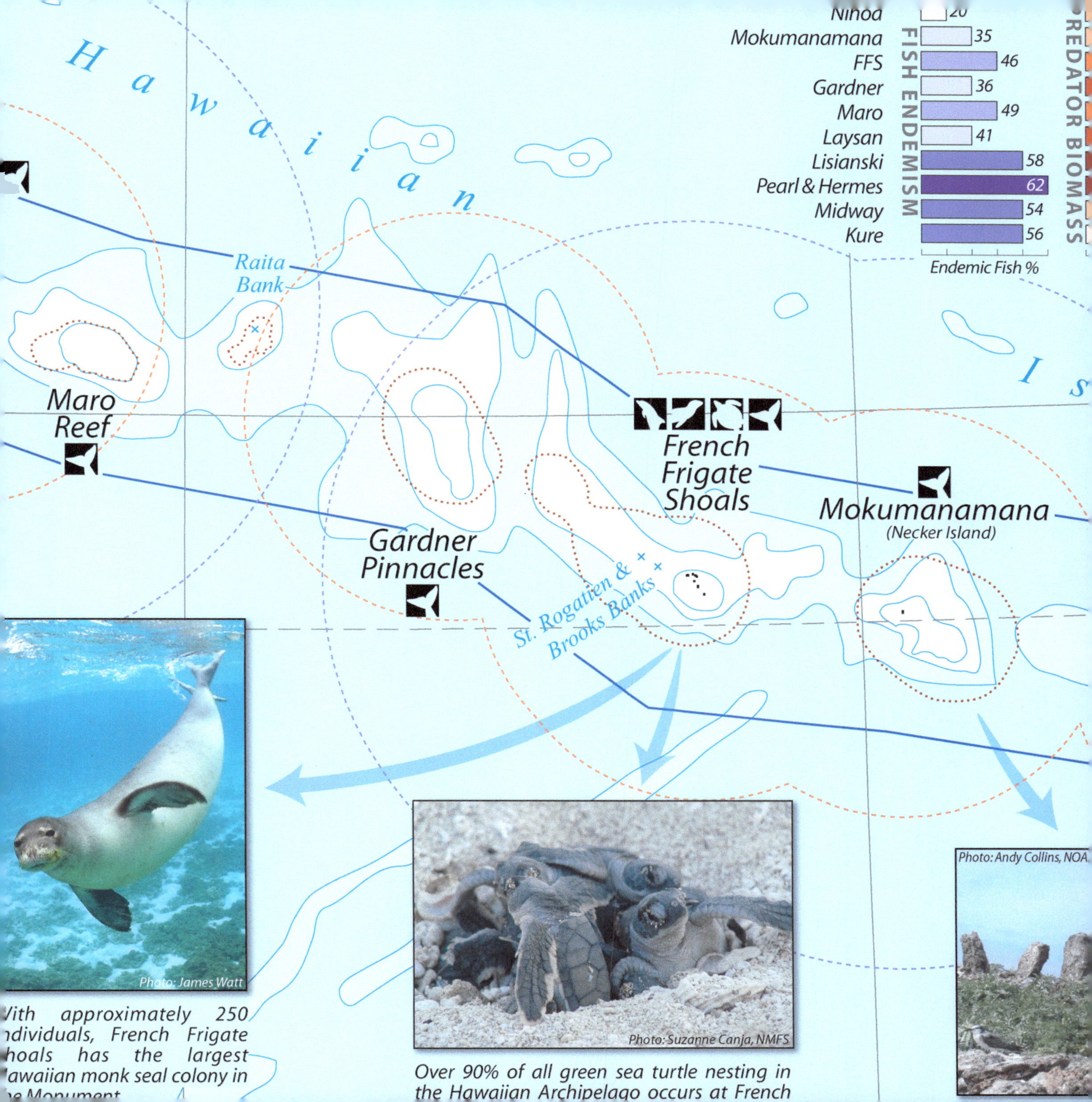

With approximately 250 individuals, French Frigate Shoals has the largest Hawaiian monk seal colony in the Monument.

Over 90% of all green sea turtle nesting in the Hawaiian Archipelago occurs at French

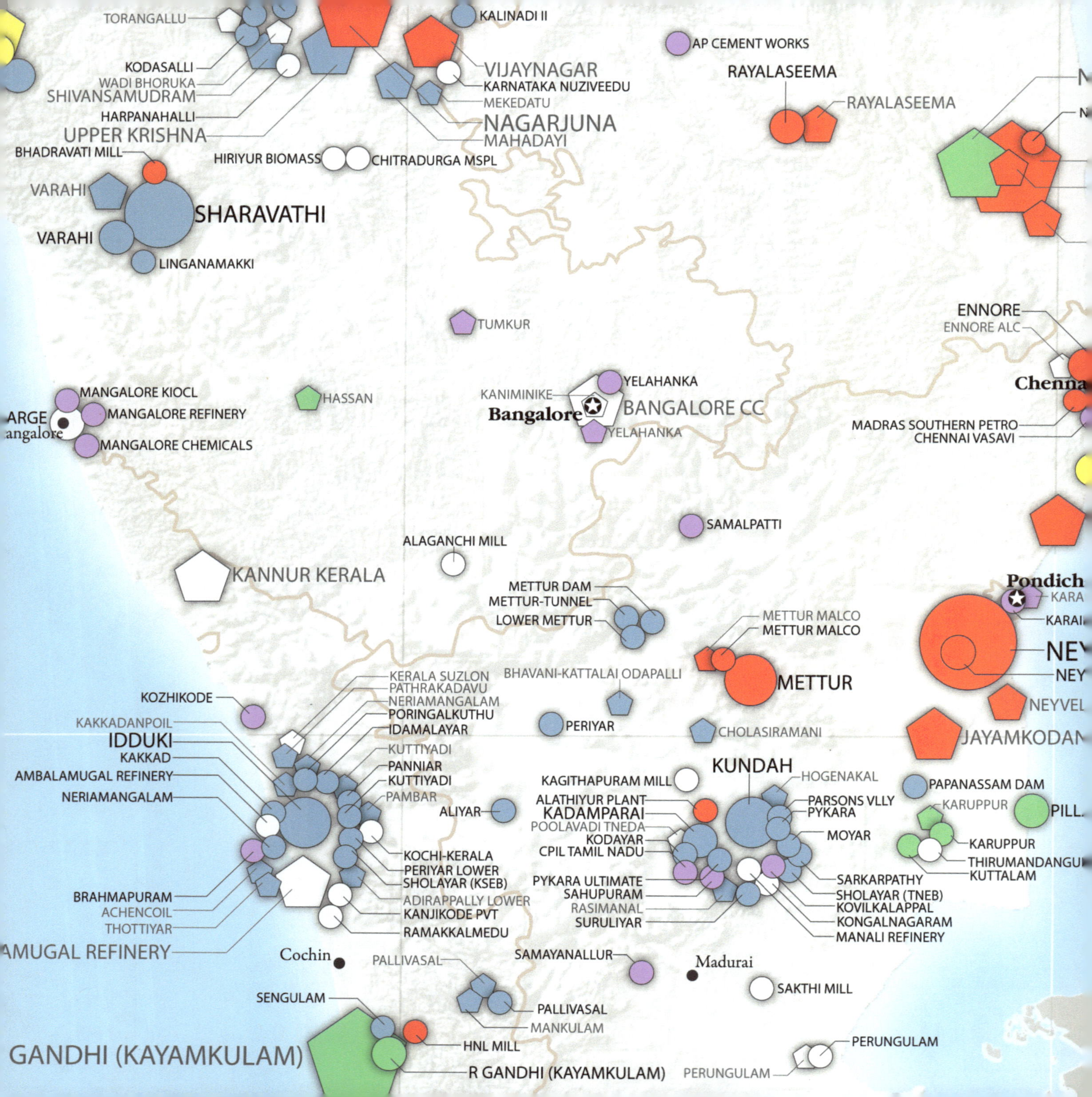

TORANGALLU
KALINADI II
AP CEMENT WORKS
KODASALLI
WADI BHORUKA
SHIVANSAMUDRAM
HARPANAHALLI
UPPER KRISHNA
VIJAYNAGAR
KARNATAKA NUZIVEEDU
MEKEDATU
NAGARJUNA
MAHADAYI
RAYALASEEMA
RAYALASEEMA
BHADRAVATI MILL
HIRIYUR BIOMASS
CHITRADURGA MSPL
VARAHI
SHARAVATHI
VARAHI
LINGANAMAKKI
TUMKUR
ENNORE
ENNORE ALC
Chenna
MANGALORE KIOCL
MANGALORE REFINERY
MANGALORE CHEMICALS
HASSAN
KANIMINIKE
YELAHANKA
Bangalore
BANGALORE CC
YELAHANKA
MADRAS SOUTHERN PETRO
CHENNAI VASAVI
SAMALPATTI
ALAGANCHI MILL
KANNUR KERALA
METTUR DAM
METTUR-TUNNEL
LOWER METTUR
Pondich
METTUR MALCO
METTUR MALCO
METTUR
KERALA SUZLON
PATHRAKADAVU
NERIAMANGALAM
PORINGALKUTHU
IDAMALAYAR
BHAVANI-KATTALAI ODAPALLI
KOZHIKODE
KAKKADANPOIL
IDDUKI
KAKKAD
AMBALAMUGAL REFINERY
NERIAMANGALAM
KUTTIYADI
PANNIAR
KUTTIYADI
PAMBAR
PERIYAR
CHOLASIRAMANI
KUNDAH
HOGENAKAL
KAGITHAPURAM MILL
PAPANASSAM DAM
ALIYAR
ALATHIYUR PLANT
KADAMPARAI
POOLAVADI TNEDA
KODAYAR
CPIL TAMIL NADU
PARSONS VLLY
PYKARA
MOYAR
KARUPPUR
KARUPPUR
THIRUMANDANGU
KUTTALAM
KOCHI-KERALA
PERIYAR LOWER
SHOLAYAR (KSEB)
ADIRAPPALLY LOWER
KANJIKODE PVT
RAMAKKALMEDU
BRAHMAPURAM
ACHENCOIL
THOTTIYAR
AMUGAL REFINERY
PYKARA ULTIMATE
SAHUPURAM
RASIMANAL
SURULIYAR
SARKARPATHY
SHOLAYAR (TNEB)
KOVILKALAPPAL
KONGALNAGARAM
MANALI REFINERY
Cochin
PALLIVASAL
SAMAYANALLUR
Madurai
SAKTHI MILL
SENGULAM
PALLIVASAL
MANKULAM
HNL MILL
GANDHI (KAYAMKULAM)
R GANDHI (KAYAMKULAM)
PERUNGULAM
PERUNGULAM

Erin John LeFevre, Claude Frank, McGraw-Hill - Platts

Power Plants of India, 2007/2008

Software used: Adobe Illustrator, Avenza MAPublisher, Adobe Photoshop

Data sources: Platts UDI and ESRI

More information: www.maps.platts.com

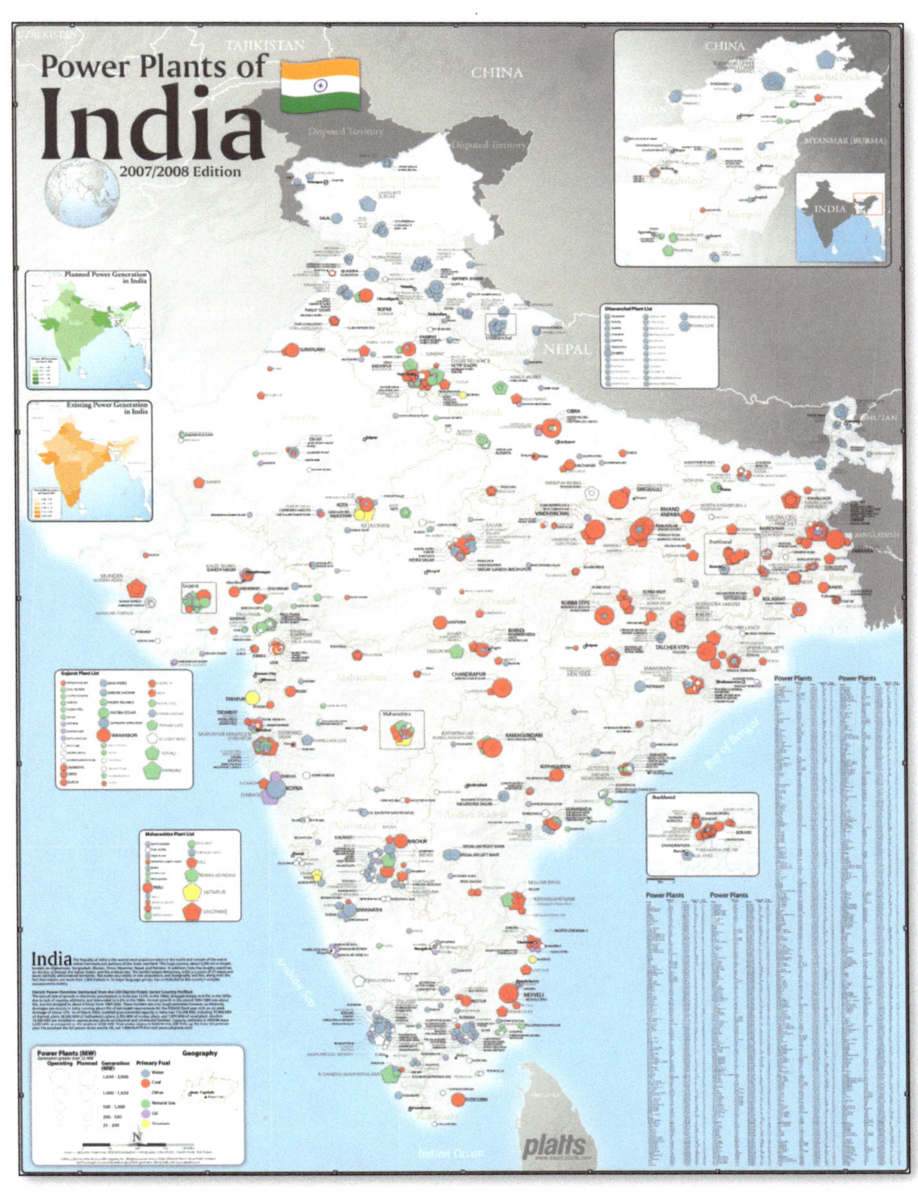

Christopher Lizotte

Three Years Later: No Child Left Behind in Dallas-Fort Worth

Software used: ESRI ArcGIS, Adobe Illustrator

Data sources: Standard and Poor's 'School Matters', Texas Education Agency, TIGER Line data

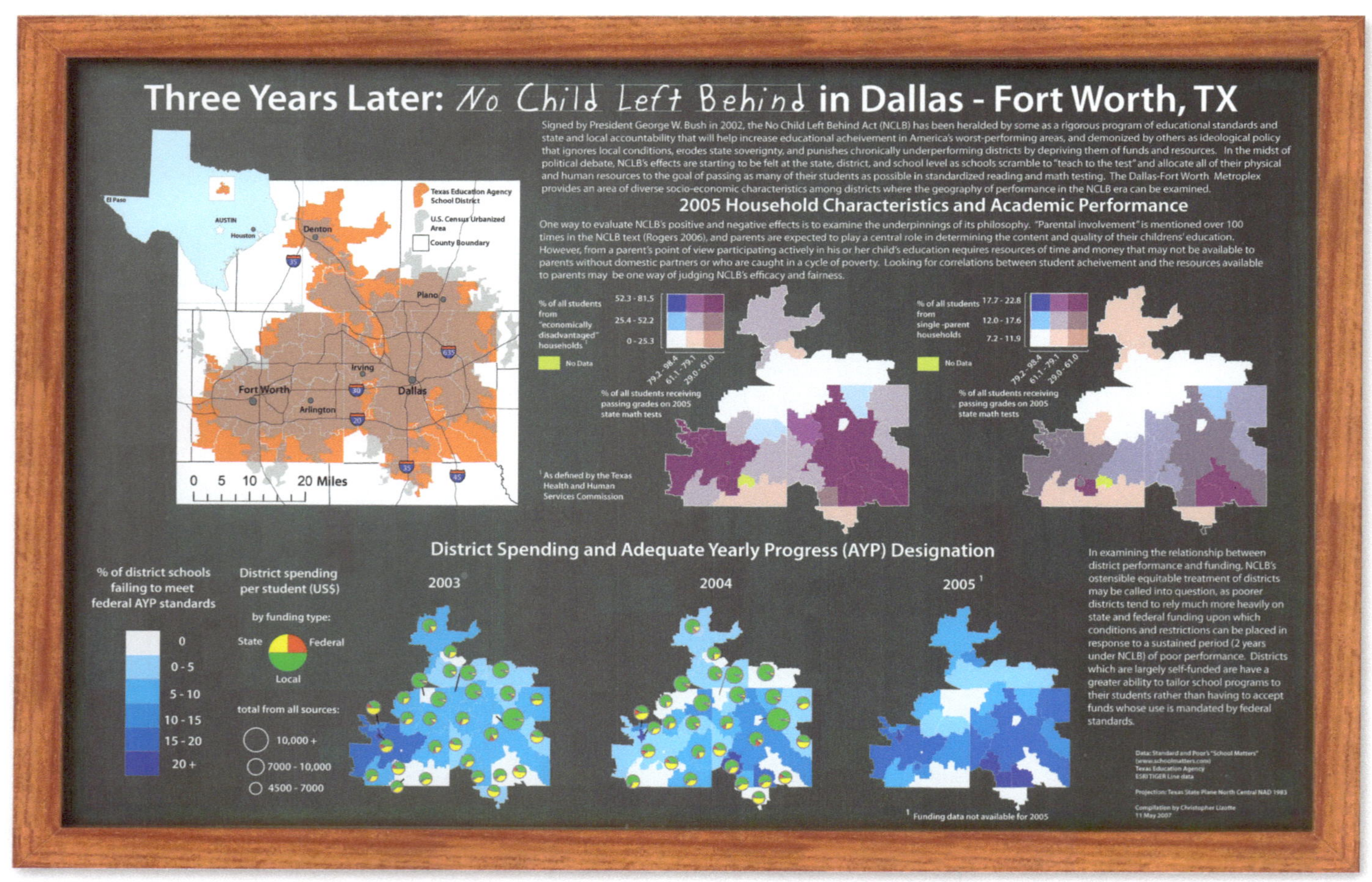

that ignores local conditions, erodes state sovereignty, and punishes chronically
political debate, NCLB's effects are starting to be felt at the state, district, and sc
and human resources to the goal of passing as many of their students as possib
provides an area of diverse socio-economic characteristics among districts whe

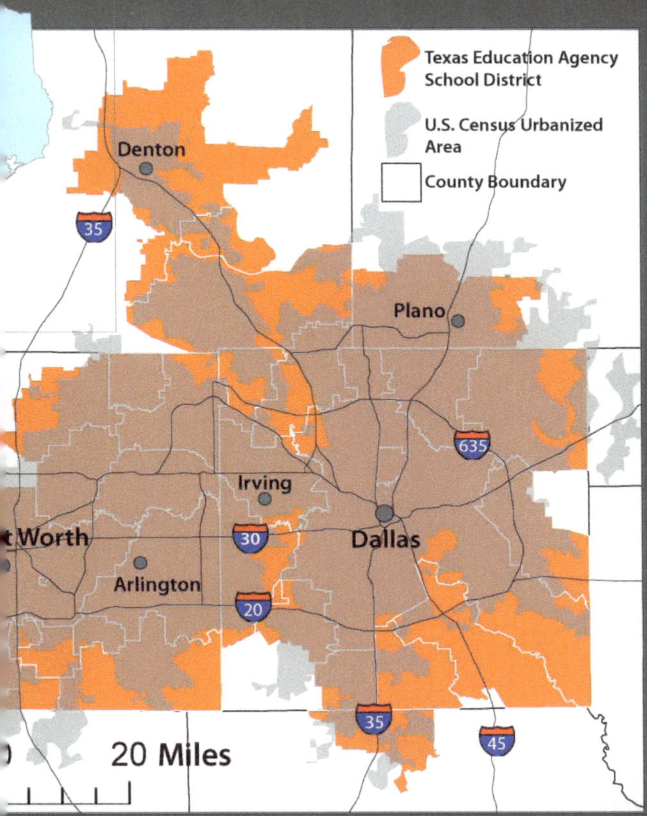

Texas Education Agency School District

U.S. Census Urbanized Area

County Boundary

Denton

Plano

Irving

Dallas

Worth

Arlington

35

635

30

20

45

20 Miles

2005 Household Characte

One way to evaluate NCLB's positive and negative effects is to examine the und
times in the NCLB text (Rogers 2006), and parents are expected to play a central
However, from a parent's point of view participating actively in his or her child's
parents without domestic partners or who are caught in a cycle of poverty. Loo
to parents may be one way of judging NCLB's efficacy and fairness.

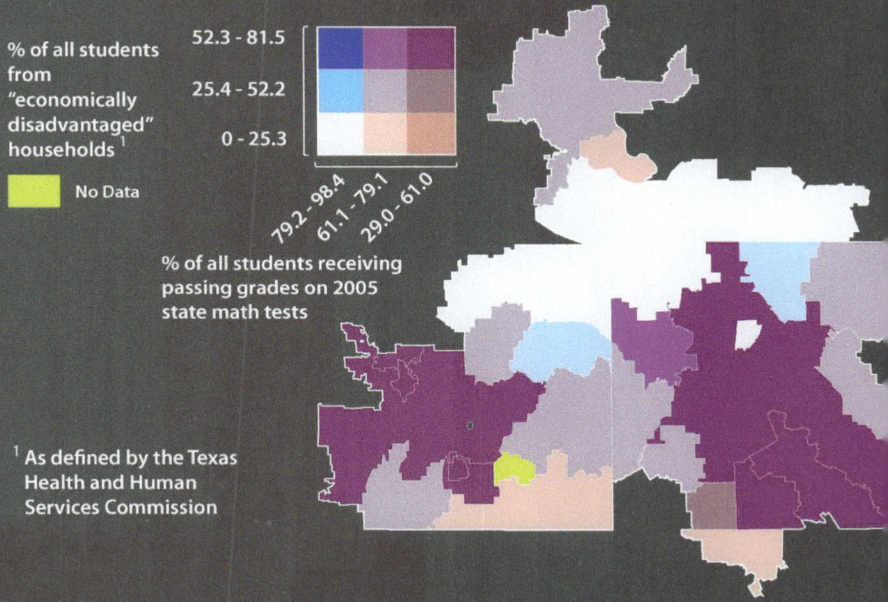

% of all students from "economically disadvantaged" households [1]

52.3 - 81.5

25.4 - 52.2

0 - 25.3

No Data

79.2 - 98.4

61.1 - 79.1

29.0 - 61.0

% of all students receiving passing grades on 2005 state math tests

[1] As defined by the Texas Health and Human Services Commission

District Spending and Adequate Yearly Progress (AYP) D

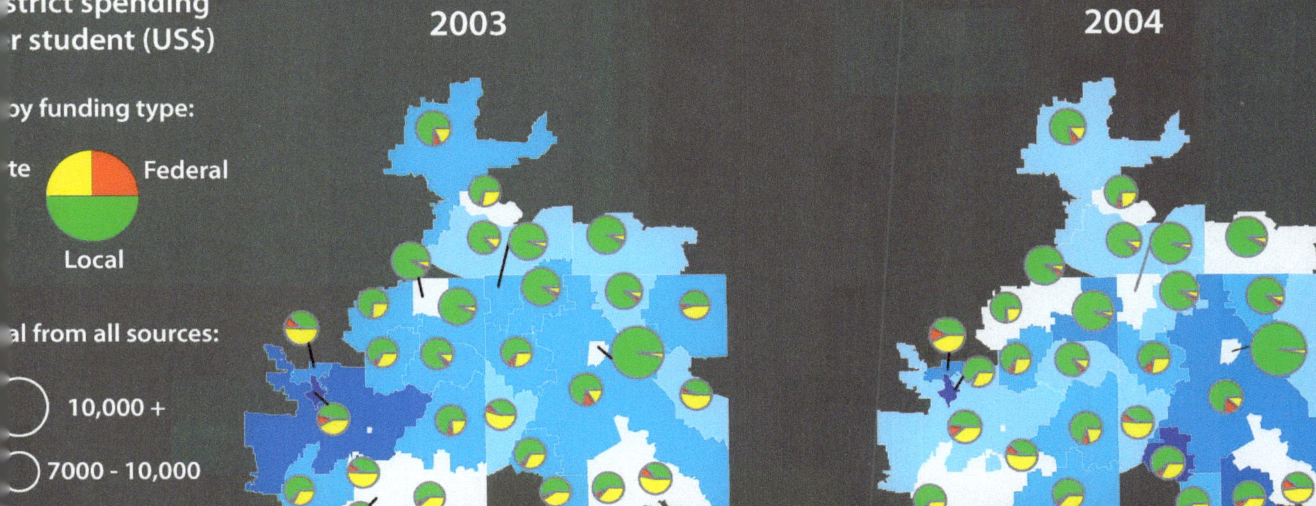

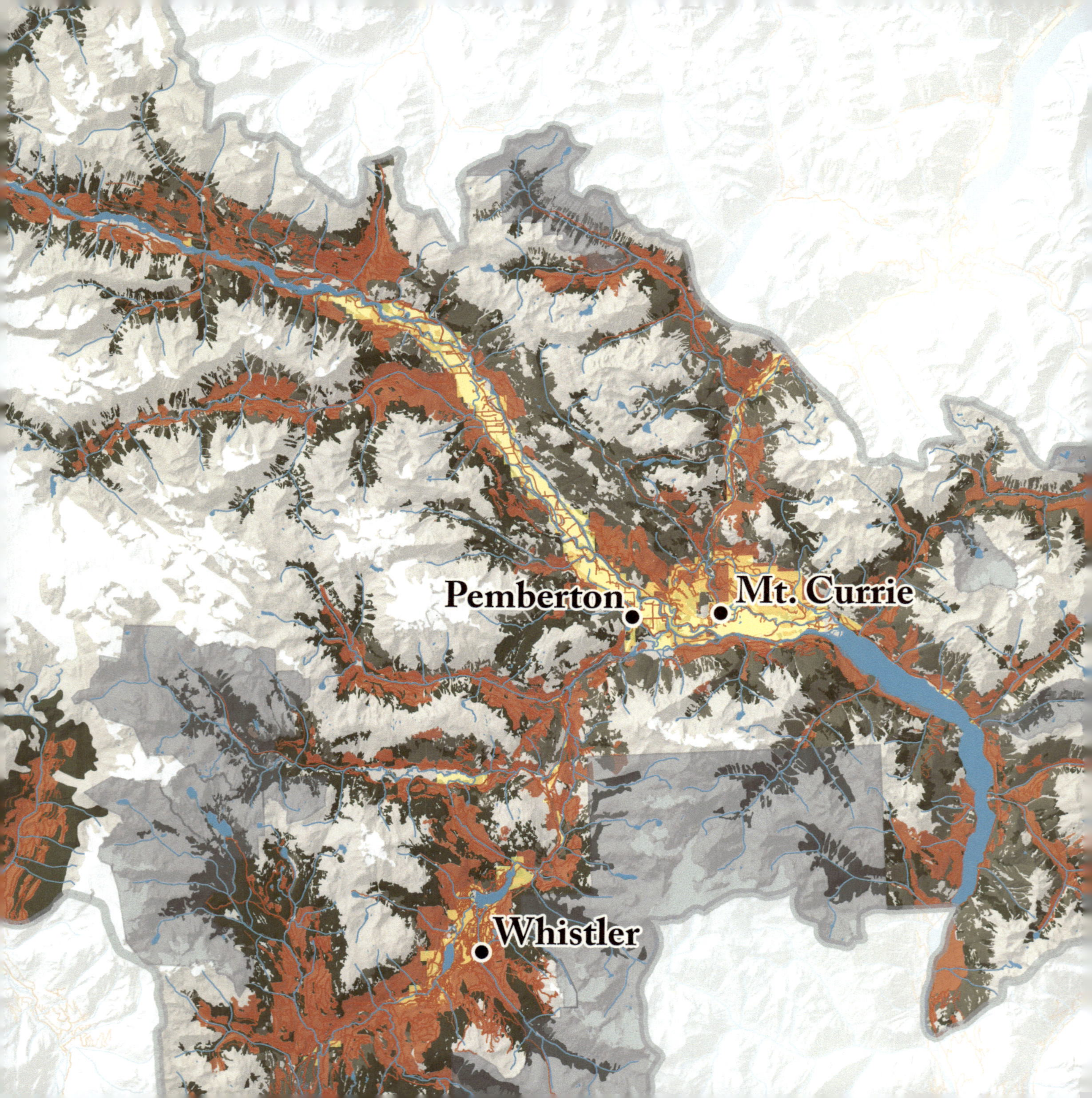

Pemberton
Mt. Currie
Whistler

Eliana Macdonald, Ecotrust Canada

Eighty Years of Landscape Change, Lil'wat Nation Traditional Territory

Software used: ESRI ArcMap, Adobe Illustrator

More information: www.ecotrust.ca

Data sources:

Baseline Thematic Mapping [shapefile]. Victoria, BC: Integrated Land Management Bureau.

Vegetation Resources Inventory [coverage]. Victoria, BC: Ministry of Forests and Range.

Young, Eric. The Soo Public Sustained Yeild Unit Record of Clearing 1920-1974 [map]. 1:125,000. In: Eric Young. The evolution of a British Columbian forest landscape, as observed in the Soo Public Sustained Yield Unit [MA thesis, Simon Fraser University, 1976].

U.S. Geological Survey (analysis by Ken Whitehead). L1-5 MSS L1G SINGLE NLAPS [col. satellite image]. Sioux Falls, S. Dak.: EROS Data Center, 1972.

U.S. Geological Survey (analysis by Ken Whitehead). L4-5 TM L1G SINGLE NLAPS [col. satellite image]. Sioux Falls, S. Dak.: EROS Data Center, 1985.

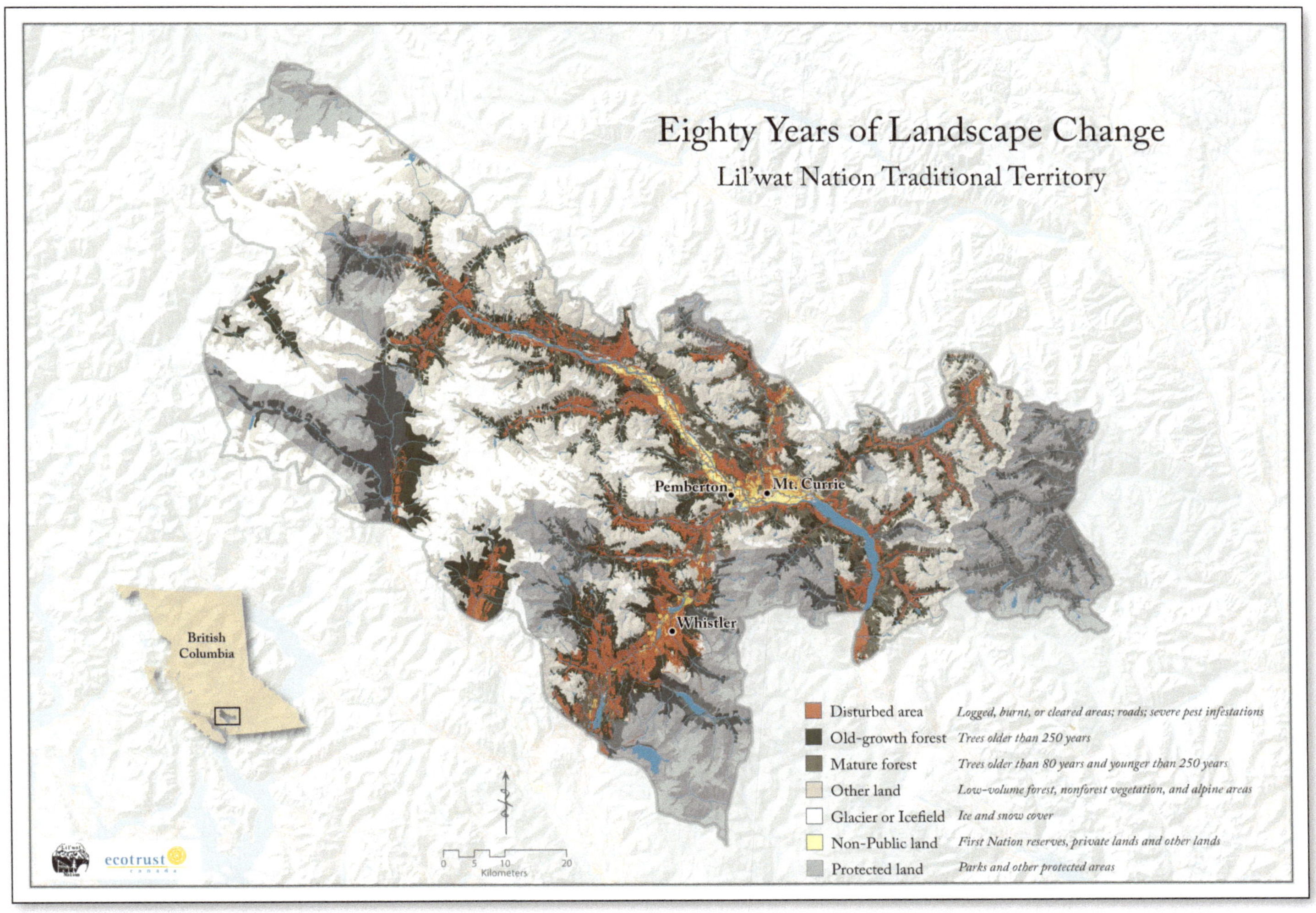

Dennis McClendon, Chicago CartoGraphics

Galesburg Bus Routes Map

Part of a folded city street map

Software used: Adobe FreeHand MX

More information: www.chicagocarto.com

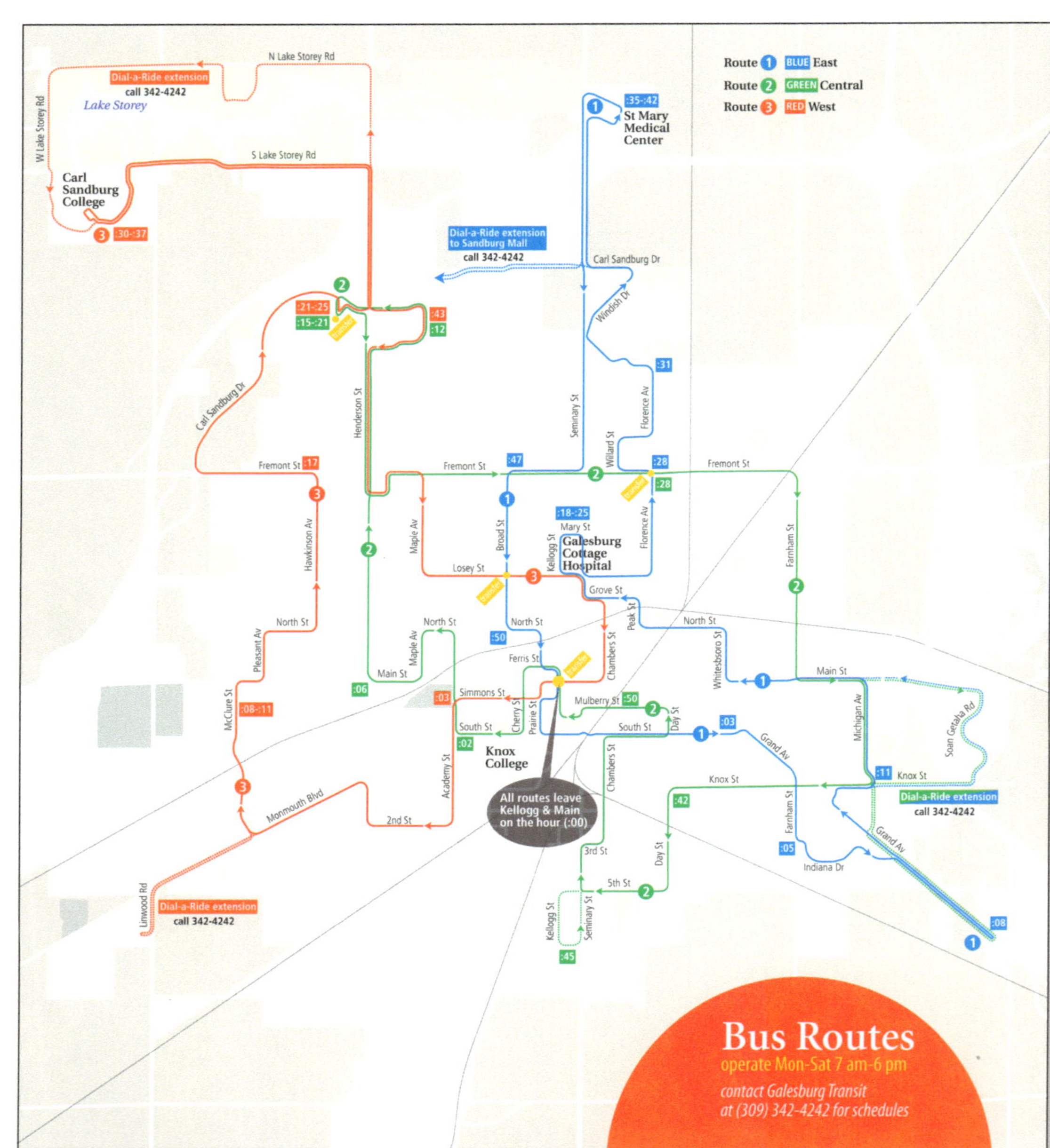

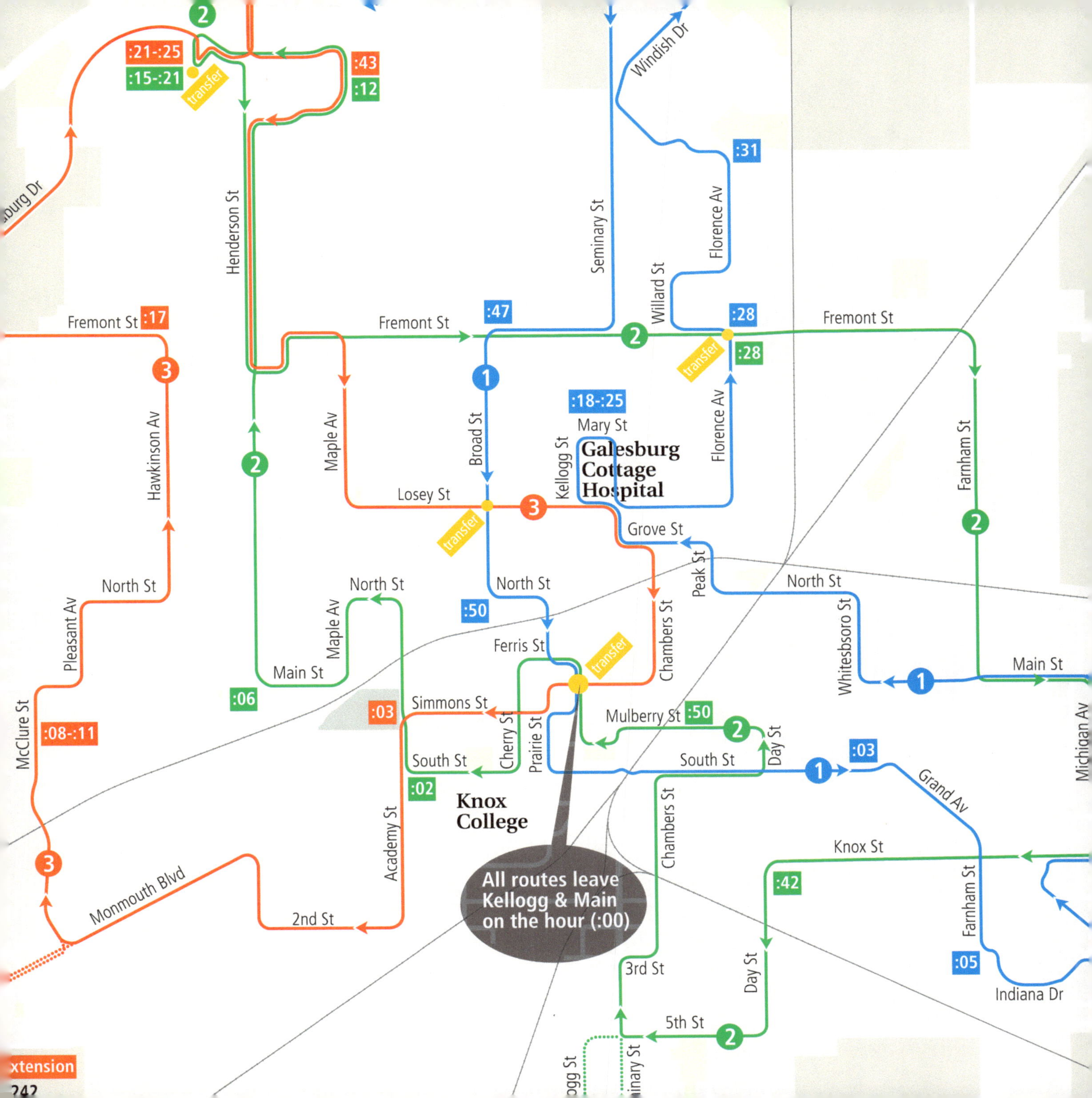

:21-:25
:15-:21
transfer
:43
:12
Windish Dr
Henderson St
Seminary St
Florence Av
:31
Willard St
Fremont St
:17
Fremont St
:47
:28
:28
Fremont St
transfer
Hawkinson Av
Maple Av
Broad St
:18-:25
Mary St
Kellogg St
Galesburg Cottage Hospital
Florence Av
Farnham St
Losey St
transfer
Grove St
Peak St
North St
North St
North St
North St
:50
Maple Av
Pleasant Av
Ferris St
transfer
Chambers St
Whitesboro St
Main St
Main St
:06
McClure St
:08-:11
:03
Simmons St
Cherry St
Prairie St
Mulberry St
:50
Day St
South St
:03
South St
:02
Knox College
Grand Av
Michigan Av
Academy St
Chambers St
Knox St
:42
Farnham St
All routes leave Kellogg & Main on the hour (:00)
Monmouth Blvd
2nd St
3rd St
Day St
:05
Indiana Dr
5th St

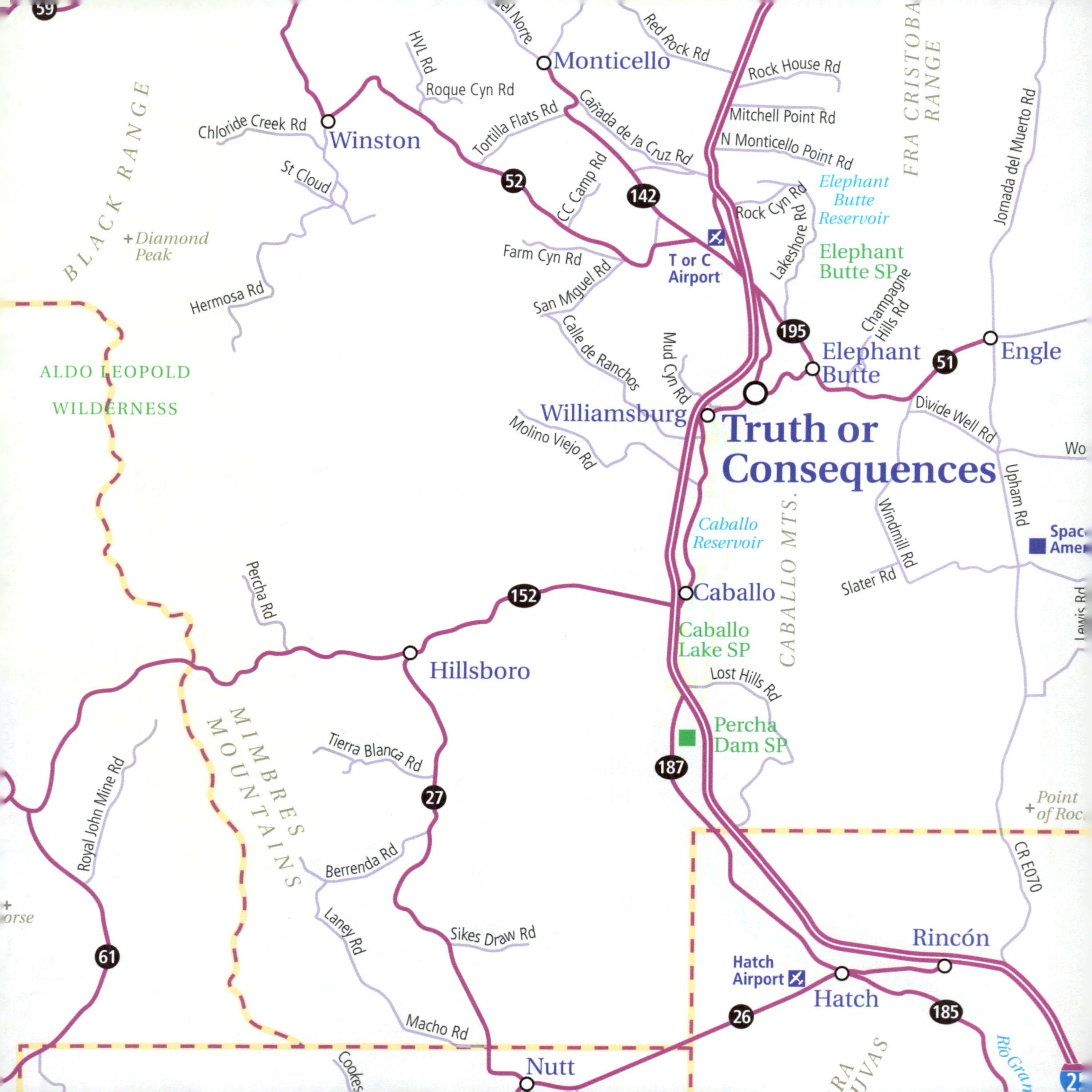

Monticello
Winston
Truth or Consequences
Williamsburg
Elephant Butte
Engle
Caballo
Hillsboro
Hatch
Rincón
Nutt
T or C Airport
Hatch Airport
Elephant Butte Reservoir
Elephant Butte SP
Caballo Reservoir
Caballo Lake SP
Percha Dam SP
BLACK RANGE
FRA CRISTOBAL RANGE
CABALLO MTS.
MIMBRES MOUNTAINS
ALDO LEOPOLD WILDERNESS
Diamond Peak
Red Rock Rd
Rock House Rd
Mitchell Point Rd
N Monticello Point Rd
HVL Rd
Roque Cyn Rd
Chloride Creek Rd
Tortilla Flats Rd
Cañada de la Cruz Rd
St Cloud
CC Camp Rd
Rock Cyn Rd
Lakeshore Rd
Farm Cyn Rd
San Miguel Rd
Calle de Ranchos
Mud Cyn Rd
Champagne Hills Rd
Jornada del Muerto Rd
Hermosa Rd
Divide Well Rd
Molino Viejo Rd
Upham Rd
Windmill Rd
Slater Rd
Percha Rd
Lost Hills Rd
Tierra Blanca Rd
Royal John Mine Rd
Berrenda Rd
Laney Rd
Sikes Draw Rd
Macho Rd
CR E070
Point of Roc
52
142
195
51
152
187
27
61
26
185

Dennis McClendon, Chicago CartoGraphics

Sierra County

Produced for Directory Plus phone book of Truth or Consequences, N.M.

Software used: Adobe FreeHand MX, MacDEM

More information: www.chicagocarto.com

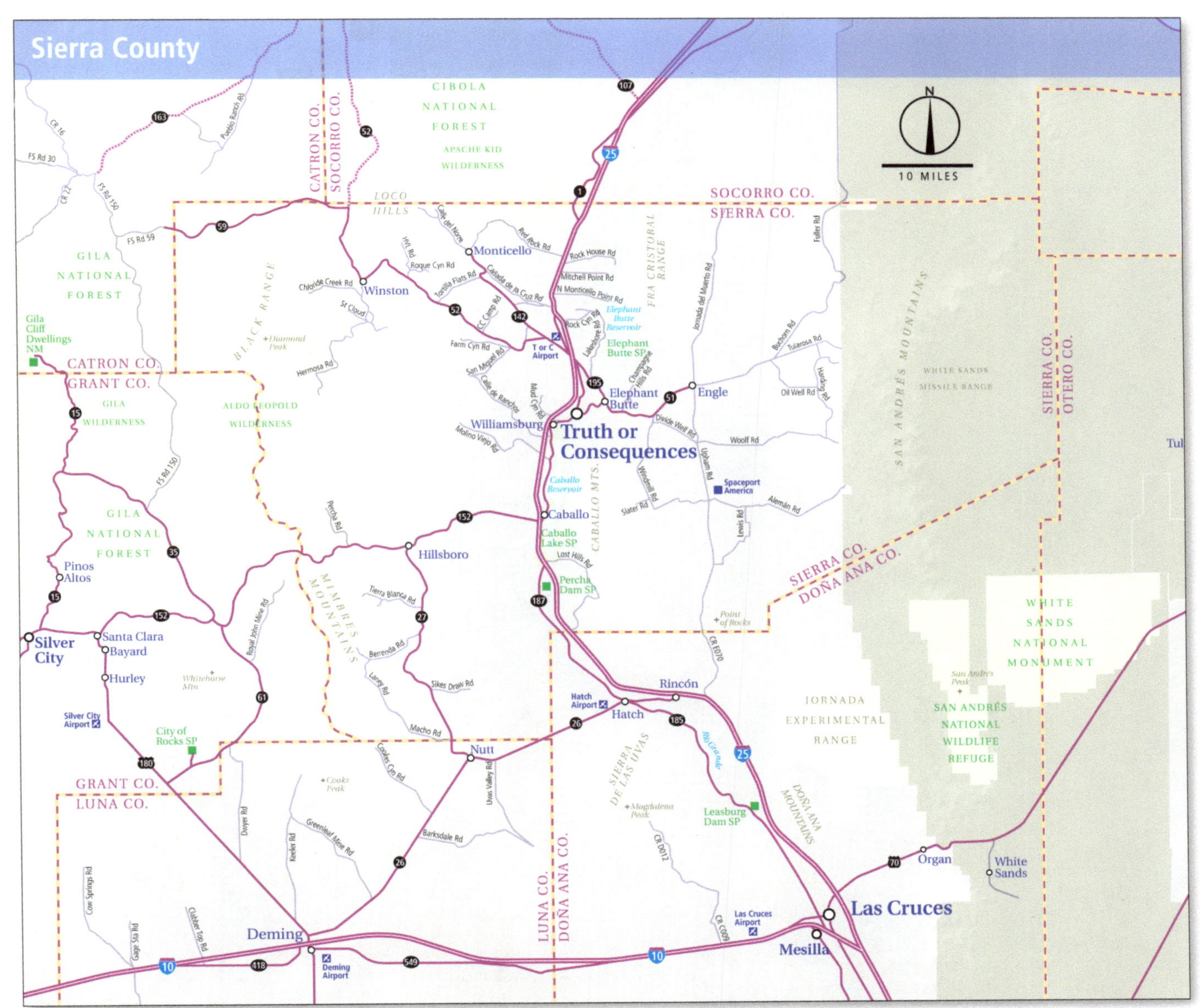

Dominik Mikiewicz, Shipping Guides Limited

The Ships Atlas - Plate 67 - Caribbean Sea

Software used: Manifold System, MicroDEM, CorelDRAW, CorelPHOTOPAINT

Data sources: Port information: Shipping Guides Limited

Vector data: WDBII, VMap0, Shipping Guides Limited

DEMs: ETOPO2, GTOPO 30, SRTM 30, SRTM 3

Other: International Maritime Organization, World Health Organization, ICC International Maritime Bureau

More information: Shipping Guides Limited: info@portinfo.co.uk

Dominik Mikiewicz: dominikmikiewicz@o2.pl, cartomatic@o2.pl

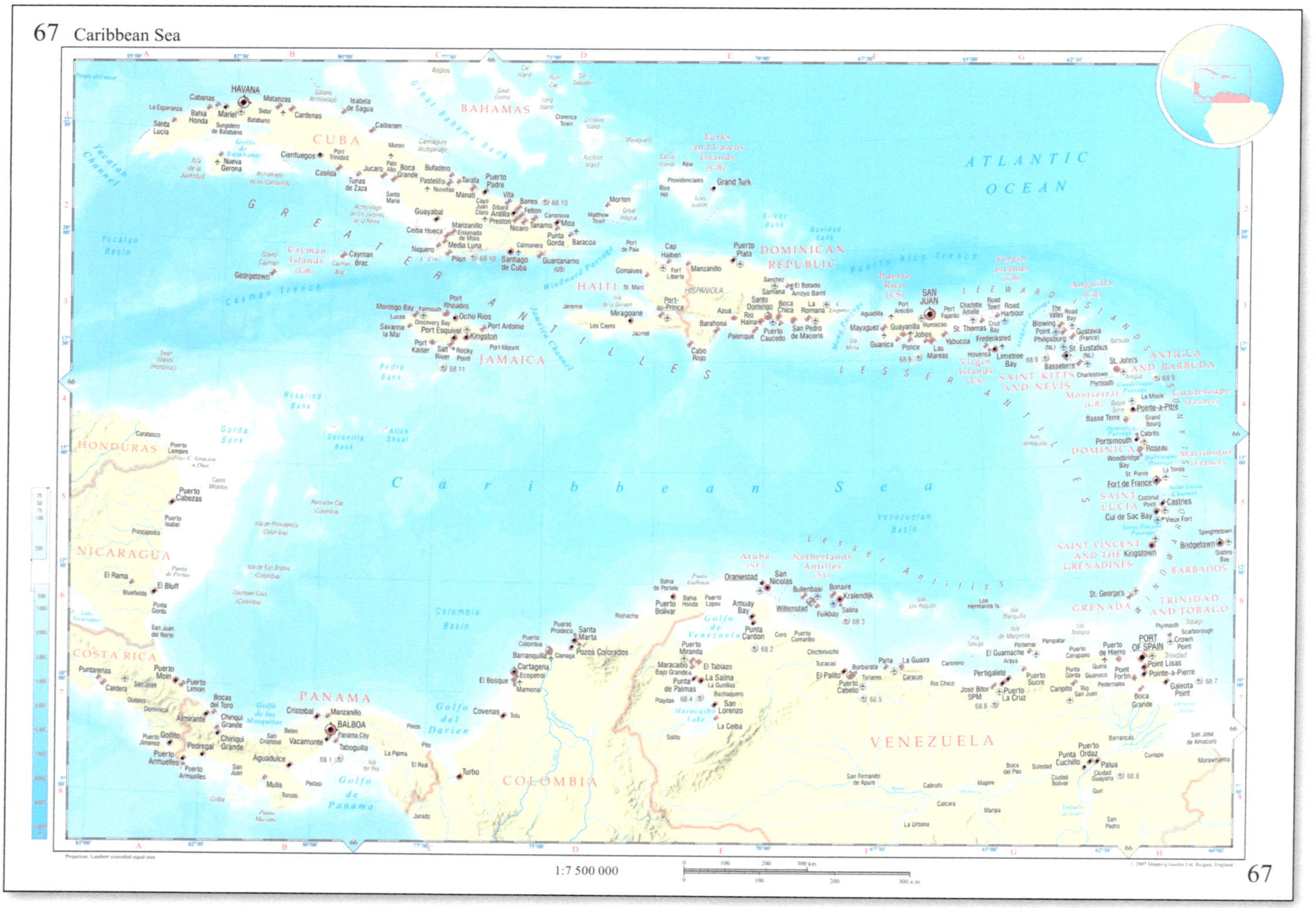

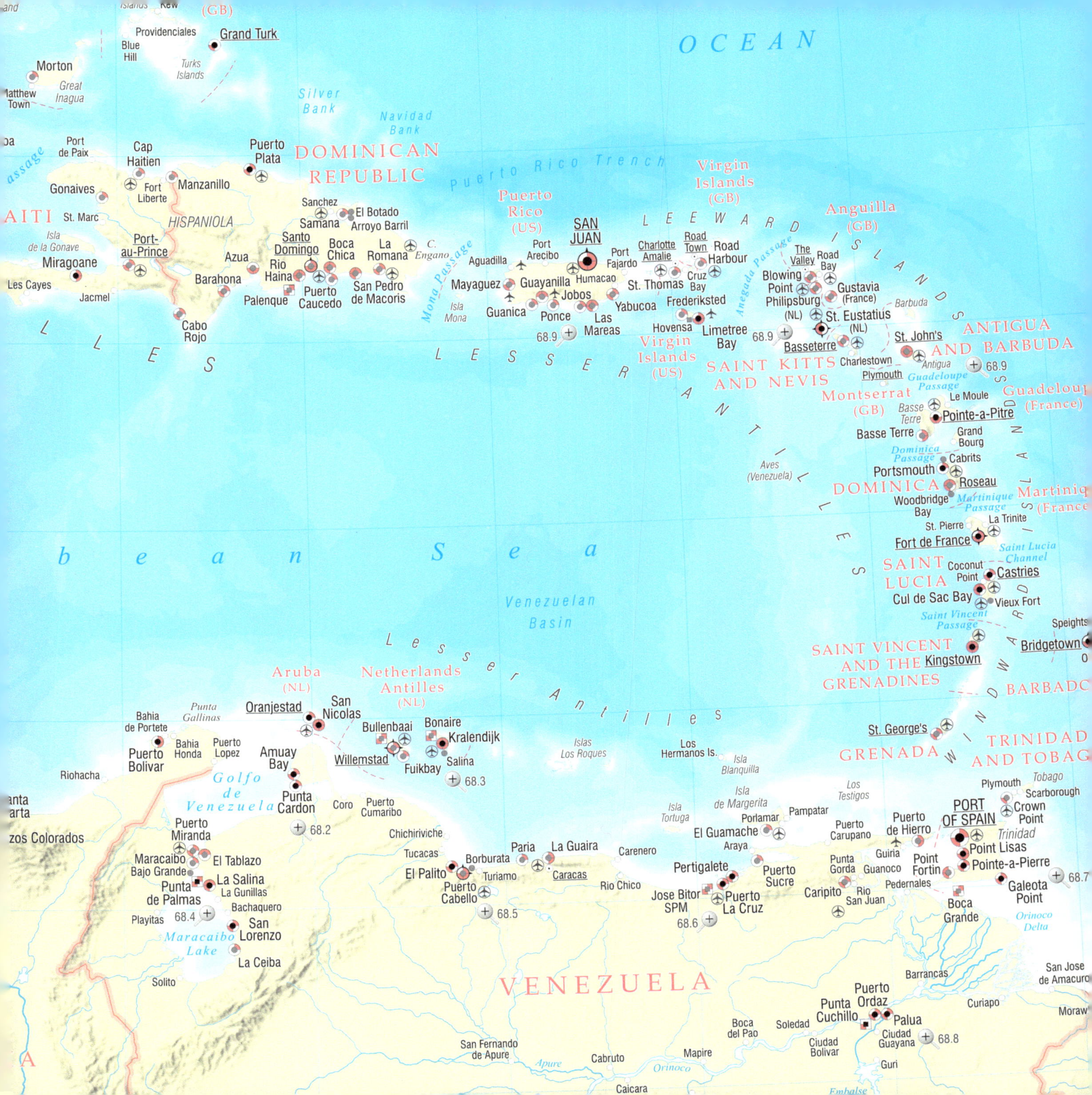

OCEAN
Providenciales
Grand Turk
Blue Hill
Turks Islands
Morton
Great Inagua
Matthew Town
Silver Bank
Navidad Bank
Port de Paix
Cap Haitien
Puerto Plata
DOMINICAN REPUBLIC
Puerto Rico Trench
Virgin Islands (GB)
Gonaives
Fort Liberte
Manzanillo
HISPANIOLA
St. Marc
Sanchez
El Botado
Samana
Arroyo Barril
Puerto Rico (US)
SAN JUAN
LEEWARD ISLANDS
Anguilla (GB)
Isla de la Gonave
Port-au-Prince
Santo Domingo
Boca Chica
La Romana
C. Engano
Aguadilla
Port Arecibo
Port Fajardo
Charlotte Amalie
Road Town
Road Harbour
The Valley
Road Bay
Miragoane
Azua
Rio Haina
Mona Passage
Mayaguez
Guayanilla
Humacao
Cruz Bay
Anegada Passage
Blowing Point
Gustavia (France)
Les Cayes
Jacmel
Barahona
Palenque
Puerto Caucedo
San Pedro de Macoris
Isla Mona
Guanica
Jobos
Ponce
Las Mareas
Yabucoa
St. Thomas
Frederiksted
Philipsburg (NL)
Barbuda
St. Eustatius (NL)
Cabo Rojo
68.9
Hovensa
Limetree Bay
Virgin Islands (US)
Basseterre
St. John's
ANTIGUA AND BARBUDA
Charlestown
Antigua
SAINT KITTS AND NEVIS
Plymouth
Guadeloupe Passage
Montserrat (GB)
Le Moule
Guadeloupe (France)
Basse Terre
Pointe-a-Pitre
Grand Bourg
Dominica Passage
Cabrits
Aves (Venezuela)
Portsmouth
DOMINICA
Roseau
Woodbridge Bay
Martinique Passage
Martinique (France)
St. Pierre
La Trinite
Fort de France
Saint Lucia Channel
LESSER ANTILLES
WINDWARD ISLANDS
bean Sea
Venezuelan Basin
SAINT LUCIA
Coconut Point
Castries
Cul de Sac Bay
Vieux Fort
Saint Vincent Passage
Speightstown
Bridgetown
SAINT VINCENT AND THE GRENADINES
Kingstown
BARBADOS
Lesser Antilles
Aruba (NL)
Netherlands Antilles (NL)
Punta Gallinas
Bahia de Portete
Oranjestad
San Nicolas
Bullenbaai
Bonaire
Kralendijk
Islas Los Roques
Los Hermanos Is.
Isla Blanquilla
St. George's
GRENADA
TRINIDAD AND TOBAGO
Puerto Bolivar
Bahia Honda
Puerto Lopez
Amuay Bay
Willemstad
Fuikbay
Salina
68.3
Riohacha
Golfo de Venezuela
Punta Cardon
Coro
Puerto Cumaribo
Isla Tortuga
Isla de Margerita
Porlamar
Pampatar
Los Testigos
Plymouth
Tobago
Scarborough
Crown Point
PORT OF SPAIN
Trinidad
Puerto Miranda
Los Colorados
68.2
Chichiriviche
El Guamache
Puerto Carupano
Puerto de Hierro
Maracaibo
El Tablazo
Bajo Grande
Tucacas
Borburata
Paria
La Guaira
Carenero
Araya
Guiria
Point Lisas
La Salina
La Gunillas
El Palito
Turiamo
Caracas
Rio Chico
Pertigalete
Puerto Sucre
Punta Gorda
Guanoco
Point Fortin
Pointe-a-Pierre
Punta de Palmas
Puerto Cabello
Jose Bitor SPM
Puerto La Cruz
Caripito
Rio San Juan
Pedernales
Galeota Point
68.7
Bachaquero
Playitas
68.4
San Lorenzo
68.5
68.6
Boca Grande
Orinoco Delta
Maracaibo Lake
La Ceiba
Solito
VENEZUELA
Barrancas
San Jose de Amacuro
Puerto Ordaz
Punta Cuchillo
Palua
Curiapo
Boca del Pao
Soledad
Ciudad Bolivar
Ciudad Guayana
68.8
San Fernando de Apure
Apure
Cabruto
Mapire
Orinoco
Guri
Caicara
Embalse

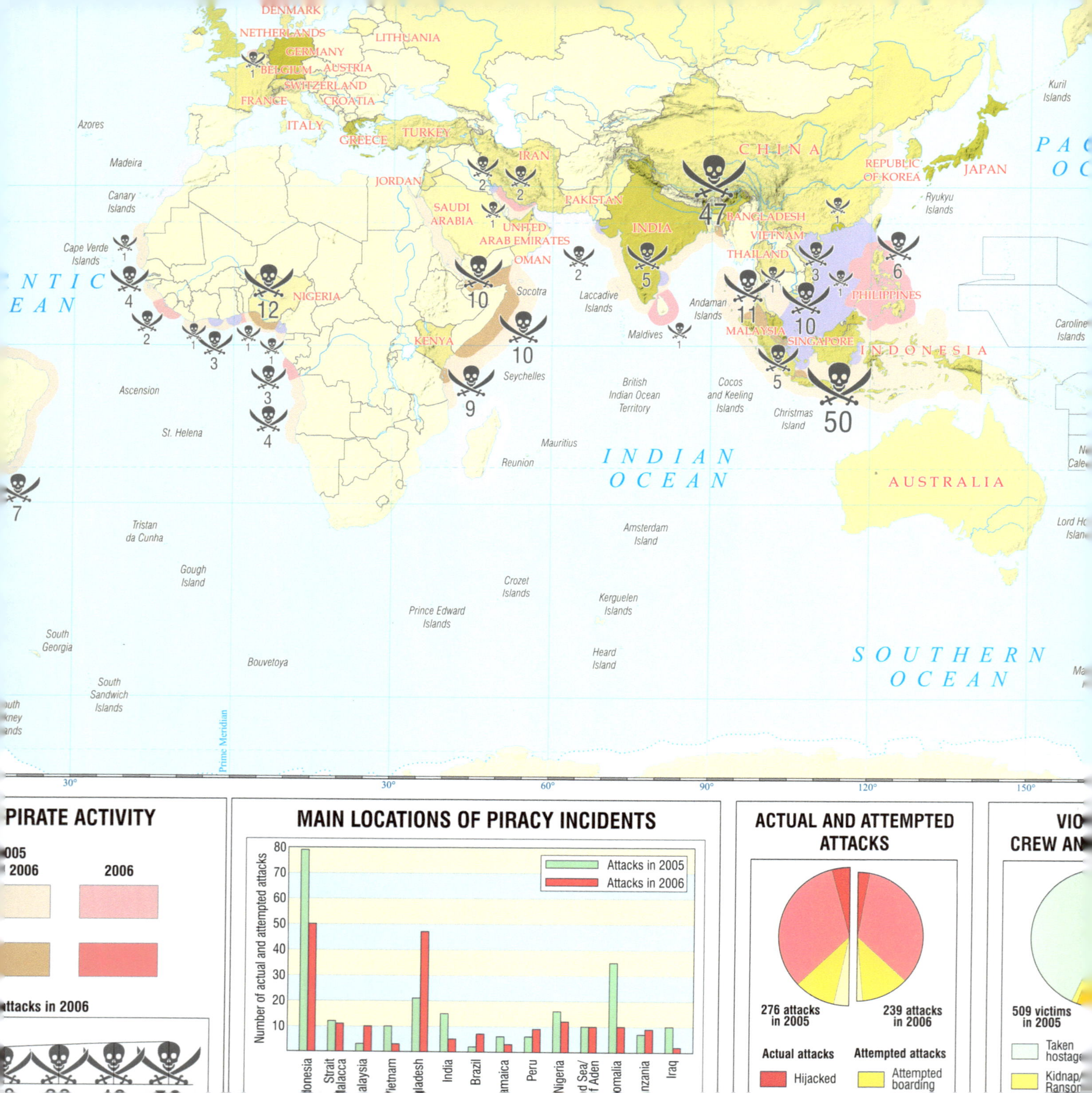

DENMARK
NETHERLANDS
LITHUANIA
GERMANY
BELGIUM
AUSTRIA
SWITZERLAND
FRANCE
CROATIA
ITALY
GREECE
TURKEY
IRAN
JORDAN
SAUDI ARABIA
UNITED ARAB EMIRATES
OMAN
PAKISTAN
INDIA
CHINA
BANGLADESH
VIETNAM
THAILAND
MALAYSIA
SINGAPORE
INDONESIA
PHILIPPINES
REPUBLIC OF KOREA
JAPAN
NIGERIA
KENYA
AUSTRALIA
Azores
Madeira
Canary Islands
Cape Verde Islands
Ascension
St. Helena
Tristan da Cunha
Gough Island
South Georgia
South Sandwich Islands
Bouvetoya
Prince Edward Islands
Crozet Islands
Kerguelen Islands
Heard Island
Amsterdam Island
Reunion
Mauritius
Seychelles
Socotra
Laccadive Islands
Maldives
Andaman Islands
British Indian Ocean Territory
Cocos and Keeling Islands
Christmas Island
Ryukyu Islands
Kuril Islands
Caroline Islands
Lord Ho Islan
INDIAN OCEAN
SOUTHERN OCEAN
NTIC EAN
PAC OC
Prime Meridian
30°
30°
60°
90°
120°
150°
1
2
2
1
1
4
2
12
1
3
1
1
3
4
7
10
10
9
2
5
47
1
11
1
3
1
10
6
1
5
50
PIRATE ACTIVITY
005
2006
2006
ttacks in 2006
MAIN LOCATIONS OF PIRACY INCIDENTS
Number of actual and attempted attacks
80
70
60
50
40
30
20
10
Attacks in 2005
Attacks in 2006
donesia
Strait alacca
alaysia
Vietnam
ladesh
India
Brazil
amaica
Peru
Nigeria
d Sea/ f Aden
omalia
nzania
Iraq
ACTUAL AND ATTEMPTED ATTACKS
276 attacks in 2005
239 attacks in 2006
Actual attacks
Attempted attacks
Hijacked
Attempted boarding
VIO CREW AN
509 victims in 2005
Taken hostage
Kidnap/ Ransom

Dominik Mikiewicz, Shipping Guides Limited

The Ships Atlas - Plate7 - Piracy Incidents

Software used: Manifold System, MicroDEM, CorelDRAW, CorelPHOTOPAINT

Data sources: Port information: Shipping Guides Limited

Vector data: WDBII, VMap0, Shipping Guides Limited

DEMs: ETOPO2, GTOPO 30, SRTM 30, SRTM 3

Other: International Maritime Organization, World Health Organization, ICC International Maritime Bureau

More information: Shipping Guides Limited: info@portinfo.co.uk

Dominik Mikiewicz: dominikmikiewicz@o2.pl, cartomatic@o2.pl

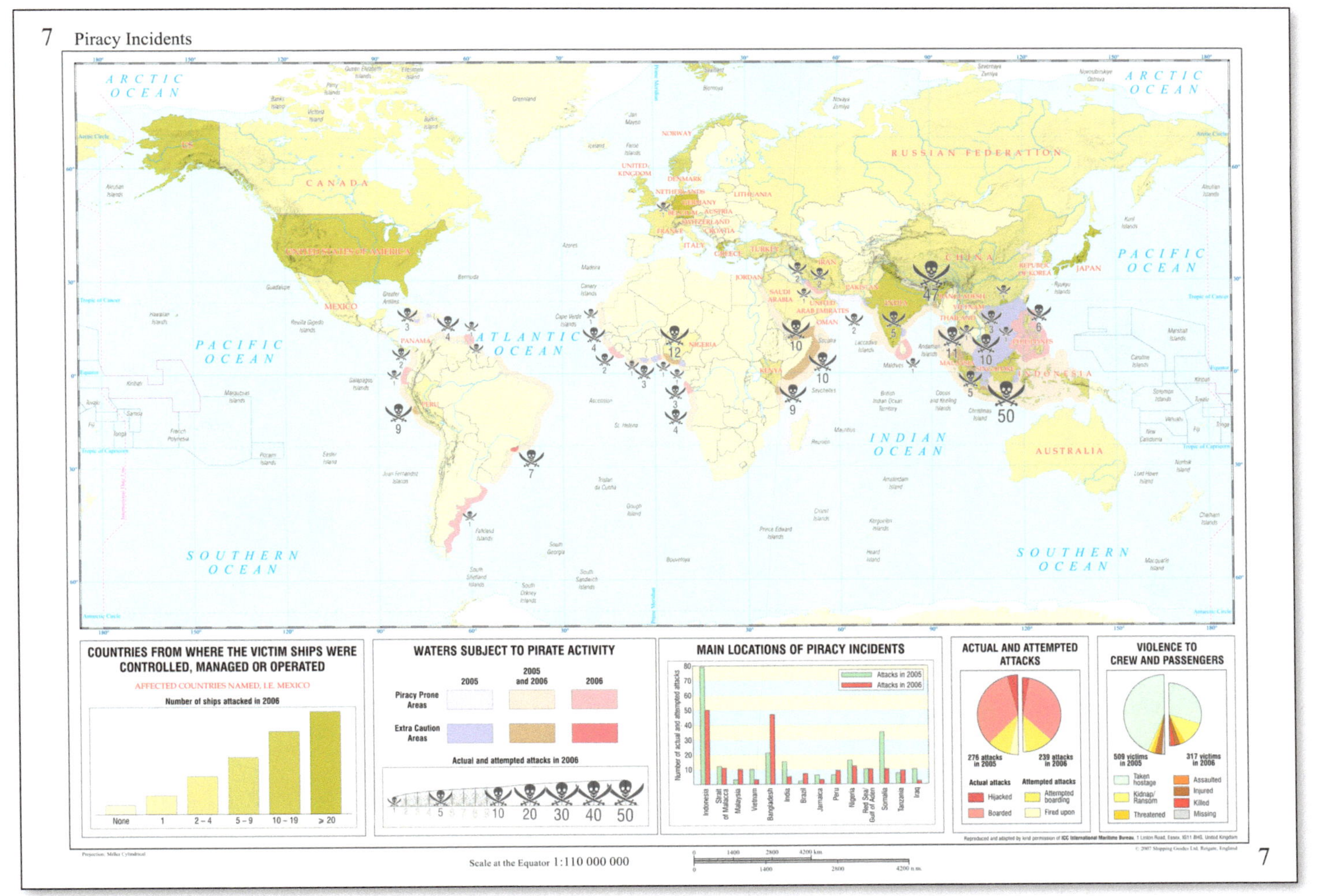

Eric Olason, Olason Cartographic Artistry

Expressive Map: Bangkok

Software used: Adobe Illustrator, Adobe Photoshop, Corel Painter

More information: www.olasoncartographic.com, www.expressivemap.com

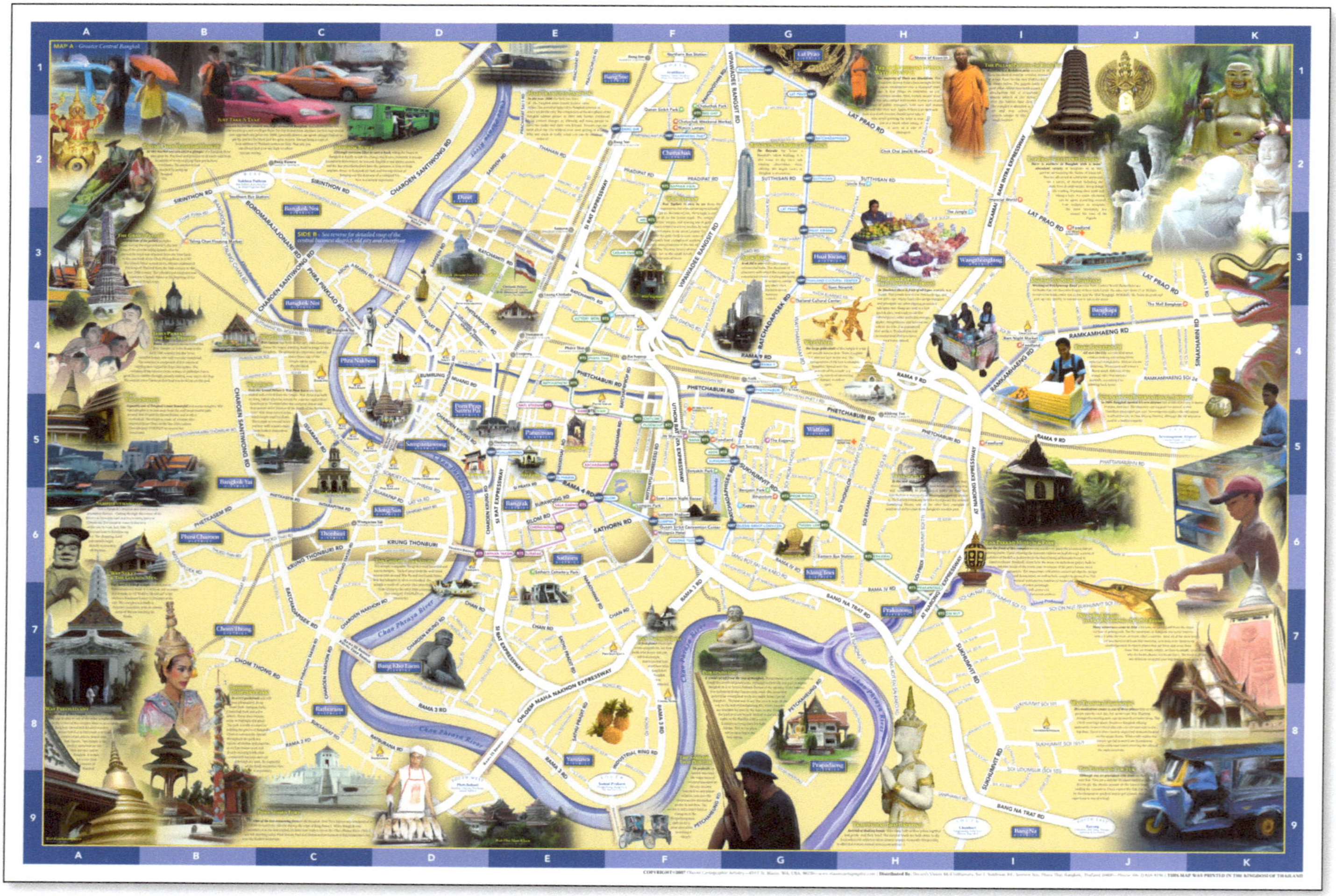

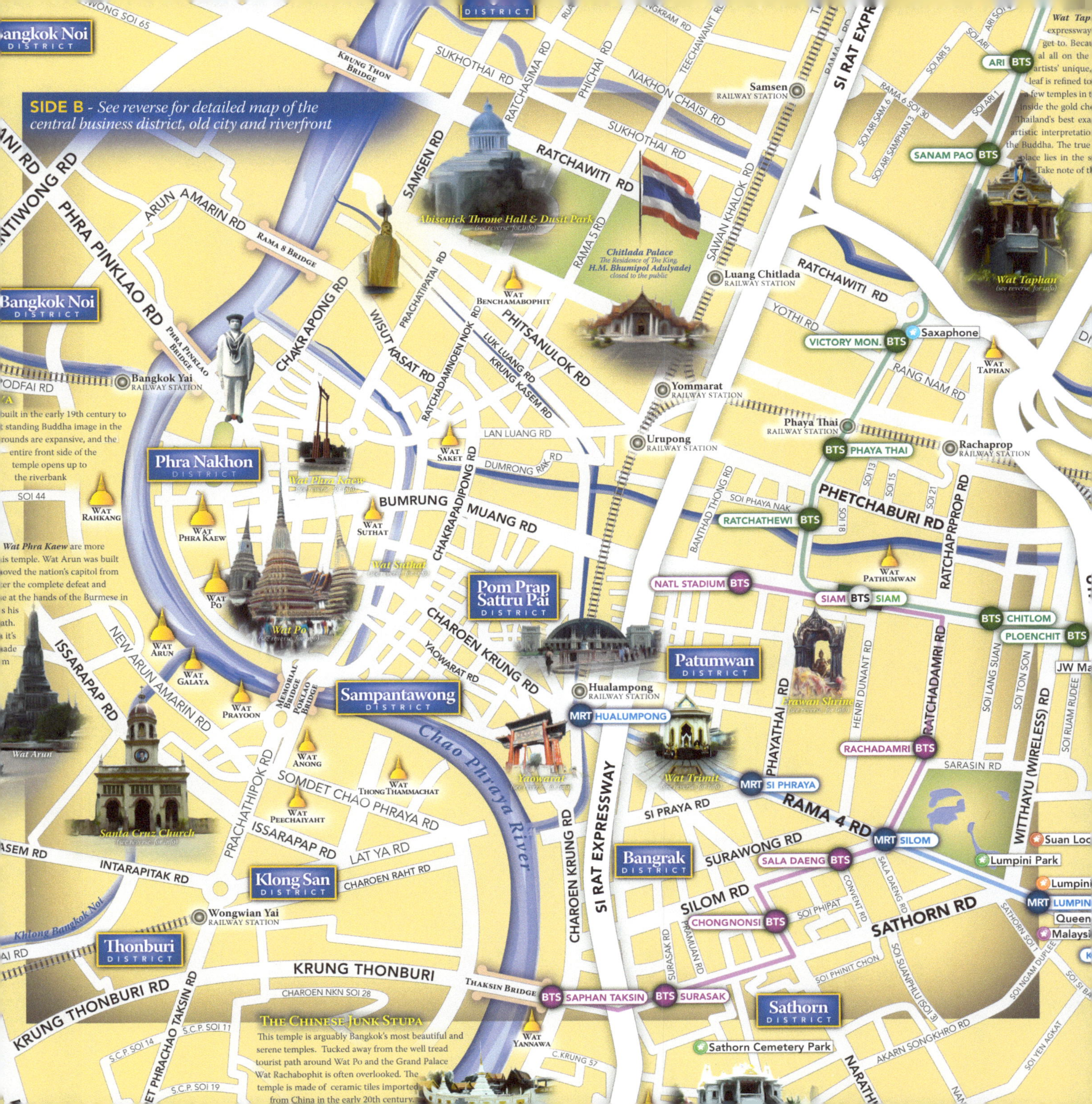
SIDE B - See reverse for detailed map of the central business district, old city and riverfront
Bangkok Noi DISTRICT
Phra Nakhon DISTRICT
Pom Prap Sattru Pai DISTRICT
Sampantawong DISTRICT
Patumwan DISTRICT
Bangrak DISTRICT
Klong San DISTRICT
Thonburi DISTRICT
Sathorn DISTRICT
Chao Phraya River
Khlong Bangkok Noi
Abisenick Throne Hall & Dusit Park
Chitlada Palace
The Residence of The King, H.M. Bhumipol Adulyadej closed to the public
Wat Taphan
Wat Phra Kaew
Wat Po
Wat Suthat
Wat Arun
Santa Cruz Church
Yaowarat
Wat Trimit
Erawan Shrine
WAT BENCHAMABOPHIT
WAT SAKET
WAT SUTHAT
WAT RAHKANG
WAT PHRA KAEW
WAT PO
WAT ARUN
WAT GALAYA
WAT PRAYOON
WAT ANONG
WAT THONG THAMMACHAT
WAT PEECHAIYAHT
WAT PATHUMWAN
WAT TAPHAN
WAT YANNAWA
KRUNG THON BRIDGE
RAMA 8 BRIDGE
PHRA PINKLAO BRIDGE
MEMORIAL BRIDGE
POKLAO BRIDGE
THAKSIN BRIDGE
Samsen RAILWAY STATION
Luang Chitlada RAILWAY STATION
Bangkok Yai RAILWAY STATION
Yommarat RAILWAY STATION
Urupong RAILWAY STATION
Phaya Thai RAILWAY STATION
Rachaprop RAILWAY STATION
Hualampong RAILWAY STATION
Wongwian Yai RAILWAY STATION
ARI BTS
SANAM PAO BTS
VICTORY MON. BTS
PHAYA THAI BTS
RATCHATHEWI BTS
NATL STADIUM BTS
SIAM BTS SIAM
CHITLOM BTS
PLOENCHIT BTS
RACHADAMRI BTS
SALA DAENG BTS
CHONGNONSI BTS
SAPHAN TAKSIN BTS
SURASAK BTS
MRT HUALUMPONG
MRT SI PHRAYA
MRT SILOM
MRT LUMPINI
Saxaphone
Lumpini Park
Sathorn Cemetery Park
Suan Loc
JW Ma
Queen
Malaysi
SUKHOTHAI RD
RATCHASIMA RD
PHICHAI RD
NAKHON CHAISI RD
TEECHAWANIT RD
SAMSEN RD
RATCHAWITI RD
SAWAN KHALOK RD
RAMA 5 RD
SI RAT EXPRESSWAY
ARUN AMARIN RD
PHRA PINKLAO RD
CHAKRAPONG RD
WISUT KASAT RD
PRACHATIPATAI RD
RATCHADAMNOEN NOK RD
LUK LUANG RD
KRUNG KASEM RD
PHITSANULOK RD
YOTHI RD
RANG NAM RD
LAN LUANG RD
DUMRONG RAK RD
BUMRUNG MUANG RD
CHAKRAPADIPONG RD
BANTHAD THONG RD
SOI PHAYA NAK
PHETCHABURI RD
RATCHAPRAROP RD
CHAROEN KRUNG RD
YAOWARAT RD
ISSARAPAP RD
NEW ARUN AMARIN RD
PRACHATHIPOK RD
SOMDET CHAO PHRAYA RD
LAT YA RD
CHAROEN RAHT RD
INTARAPITAK RD
KRUNG THONBURI RD
KRUNG THONBURI
CHAROEN NKN SOI 28
PHRACHAO TAKSIN RD
S.C.P. SOI 11
S.C.P. SOI 14
S.C.P. SOI 19
C.KRUNG 57
PHAYATHAI RD
HENRI DUNANT RD
RATCHADAMRI RD
SOI LANG SUAN
SOI TON SON
WITTHAYU (WIRELESS) RD
SOI RUAM RUDEE
SARASIN RD
RAMA 4 RD
SI PRAYA RD
SURAWONG RD
SILOM RD
SOI PHIPAT
CONVENT RD
SALA DAENG RD
SATHORN RD
SOI PHINIT CHON
SOI SUANPHLU (SOI 3)
AKARN SONGKHRO RD
SURASAK RD
PRAMUAN RD
SOI NGAM DUPLEE
SATHORN SOI 1
SOI YEN AGKAT
SOI 44
SOI 13
SOI 15
SOI 18
SOI 21
SOI ARI SAM 6
SOI ARI SAMPHAN 3
RAMA 6 SOI 30
SOI ARI 5
...built in the early 19th century to ...t standing Buddha image in the ...rounds are expansive, and the ...entire front side of the temple opens up to the riverbank
Wat Phra Kaew are more ...is temple. Wat Arun was built ...moved the nation's capitol from ...ter the complete defeat and ...e at the hands of the Burmese in
Wat Tap... expressway ... get to. Becau... al all on the ... artists' unique, ... leaf is refined to ... few temples in t... inside the gold ch... Thailand's best exa... artistic interpretatio... the Buddha. The true ... place lies in the s... Take note of th...
THE CHINESE JUNK STUPA
This temple is arguably Bangkok's most beautiful and serene temples. Tucked away from the well tread tourist path around Wat Po and the Grand Palace Wat Rachabophit is often overlooked. The temple is made of ceramic tiles imported from China in the early 20th century.

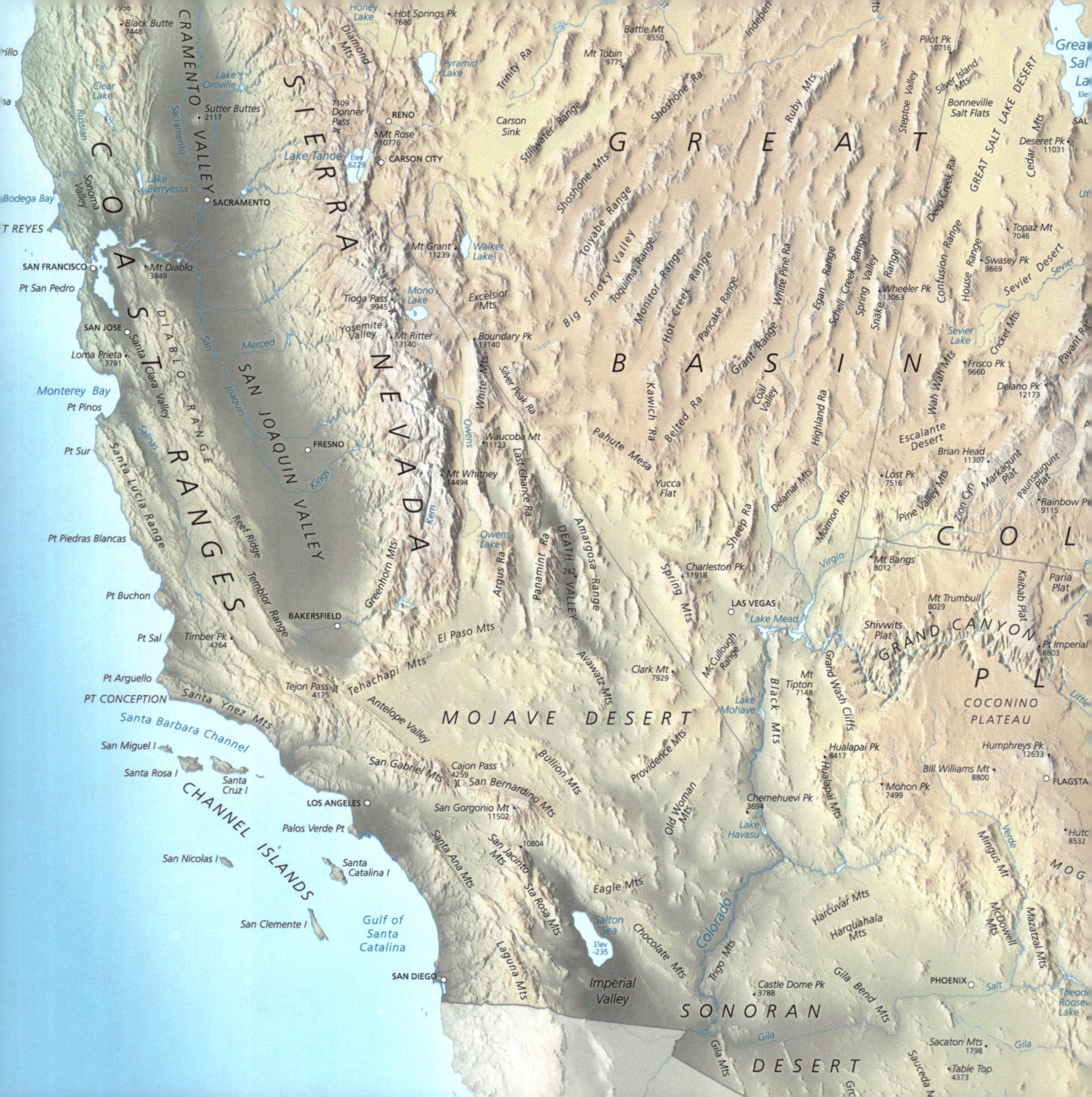

Black Butte
7448
CRAMENTO VALLEY
Honey Lake
Hot Springs Pk
7680
Diamond Mts
Pyramid Lake
Battle Mt
8550
Indepen
Pilot Pk
10716
Great Salt La
Lake Oroville
Clear Lake
Trinity Ra
Mt Tobin
9775
Silver Island Mts
Bonneville Salt Flats
GREAT SALT LAKE DESERT
Steptoe Valley
Russian
Sutter Buttes
2117
SIERRA
7109
Donner Pass
RENO
Mt Rose
10776
Carson Sink
Stillwater Range
Shoshone Ra
Ruby Mts
GREAT
Deseret Pk
11031
Cedar Mts
COAST RANGES
Sacramento
Lake Tahoe
Elev 6229
CARSON CITY
Shoshone Mts
Deep Creek Ra
Sonoma Valley
Lake Berryessa
SACRAMENTO
Bodega Bay
Toiyabe Range
T REYES
Topaz Mt
7046
Walker Lake
Mt Grant
11239
Big Smoky Valley
Toquima Range
Monitor Range
Hot Creek Range
Pancake Range
White Pine Ra
Egan Range
Schell Creek Range
Spring Valley
Snake Range
Confusion Range
House Range
Swasey Pk
9669
Sevier Desert
Sevier
SAN FRANCISCO
Mt Diablo
3849
Pt San Pedro
Tioga Pass
9945
Mono Lake
Excelsior Mts
Wheeler Pk
13063
SAN JOSE
DIABLO RANGE
Yosemite Valley
Mt Ritter
13140
Boundary Pk
13140
Sevier Lake
Cricket Mts
Loma Prieta
3791
Santa Clara Valley
Merced
San
NEVADA
White Mts
Silver Peak Ra
BASIN
Grant Range
Coal Valley
Frisco Pk
9660
Pavant Ra
Monterey Bay
Pt Pinos
SAN JOAQUIN VALLEY
Joaquin
Kawich Ra
Belted Ra
Highland Ra
Wah Wah Mts
Delano Pk
12173
Salinas
FRESNO
Owens
Waucoba Mt
11123
Pahute Mesa
Escalante Desert
Brian Head
11307
Pt Sur
Santa Lucia Range
Kings
Mt Whitney
14494
Last Chance Ra
Yucca Flat
Delamar Mts
Lost Pk
7516
Markagunt Plat
Paunsaugunt Plat
Rainbow P
9115
Kern
Mormon Mts
Pine Valley Mts
Zion Cyn
Owens Lake
Panamint Ra
DEATH VALLEY
-282
Amargosa Range
Sheep Ra
Pt Piedras Blancas
Reef Ridge
COL
Greenhorn Mts
Argus Ra
Spring Mts
Charleston Pk
11918
Virgin
Mt Bangs
8012
Kaibab Plat
Paria Plat
Temblor Range
Pt Buchon
Mt Trumbull
8029
LAS VEGAS
Lake Mead
BAKERSFIELD
Shivwits Plat
GRAND CANYON
Pt Imperial
8803
El Paso Mts
Pt Sal
Timber Pk
4764
Avawatz Mts
McCullough Range
Clark Mt
7929
Mt Tipton
7148
Grand Wash Cliffs
PL
Pt Arguello
Tejon Pass
4175
Tehachapi Mts
PT CONCEPTION
Antelope Valley
MOJAVE DESERT
Lake Mohave
Black Mts
COCONINO PLATEAU
Little
Santa Ynez Mts
Santa Barbara Channel
Providence Mts
San Miguel I
Bullion Mts
Hualapai Pk
8417
Humphreys Pk
12633
Santa Rosa I
Santa Cruz I
San Gabriel Mts
Cajon Pass
4259
San Bernardino Mts
Hualapai Mts
Bill Williams Mt
8800
FLAGSTA
CHANNEL ISLANDS
Chemehuevi Pk
3694
Mohon Pk
7499
LOS ANGELES
San Gorgonio Mt
11502
Old Woman Mts
Palos Verde Pt
Santa Ana Mts
San Jacinto Mts
10804
Lake Havasu
Verde
Hutc
8532
San Nicolas I
Santa Catalina I
Mingus Mt
MOG
Eagle Mts
Sta Rosa Mts
Harcuvar Mts
Colorado
San Clemente I
Gulf of Santa Catalina
Salton Sea
Harquahala Mts
McDowell Mts
Mazatzal Mts
Elev -235
Chocolate Mts
Trigo Mts
SAN DIEGO
Laguna Mts
Imperial Valley
Castle Dome Pk
3788
Gila Bend Mts
PHOENIX
Salt
Theodo Roose Lake
SONORAN
Gila
Sacaton Mts
1798
Gila
Gila Mts
DESERT
Table Top
4373
Sauceda

Tom Patterson, U.S. National Park Service

Physical map of the Coterminous United States

In the public domain

Software used: Adobe Illustrator, Adobe Photoshop, Natural Scene Designer, Avenza MAPublisher, Cartagena, Graffix

Data sources: Digital Chart of the World, GTOPO30, SRTM, CleanTOPO2

More information: www.shadedrelief.com

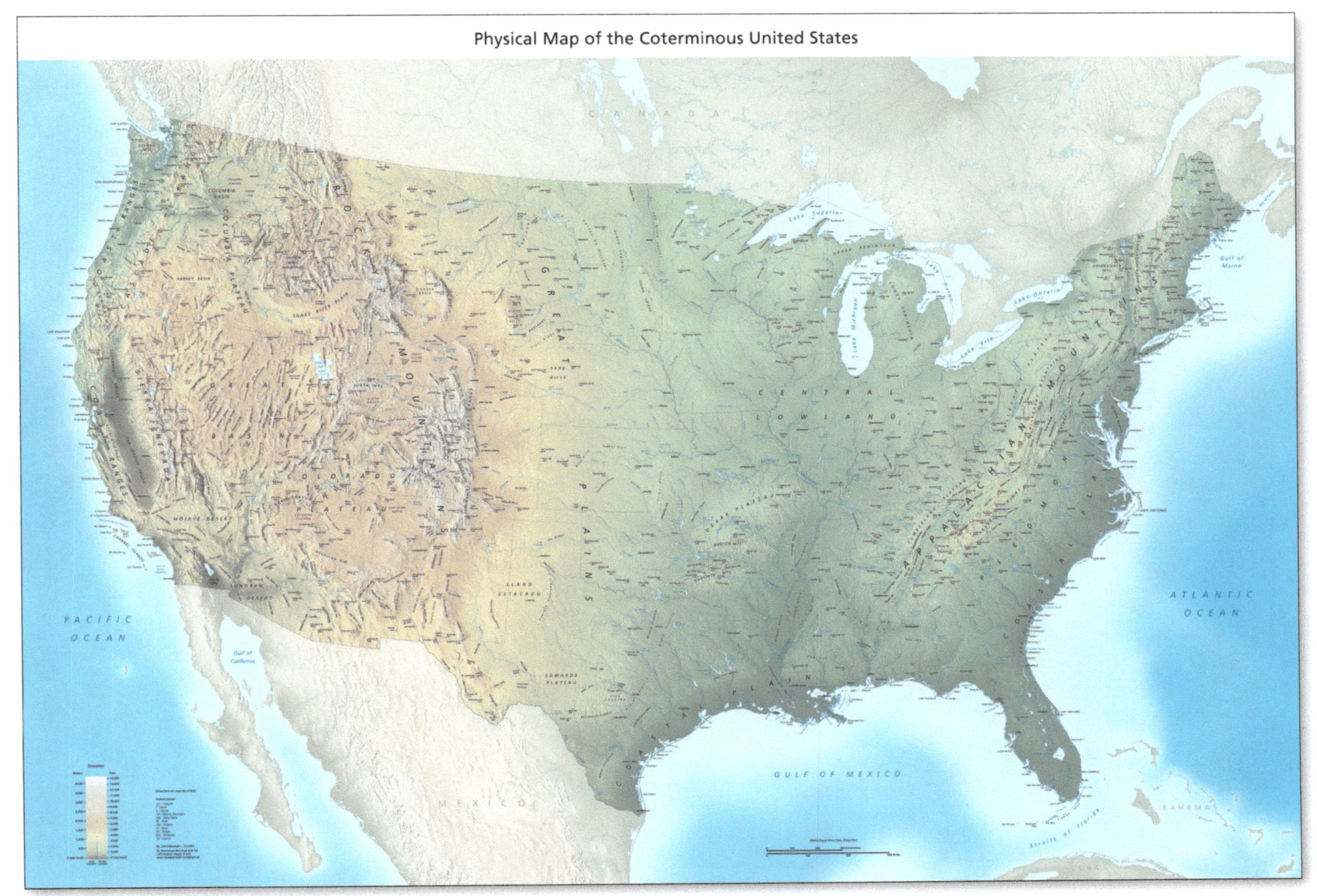

Jean-Louis Rheault

Okanagan Valley Geopictorial™ Map
For Okanagan Map Guides

Media used: Ink, Watercolour, AdobePhotoshop
More information: jl@geografix.ca

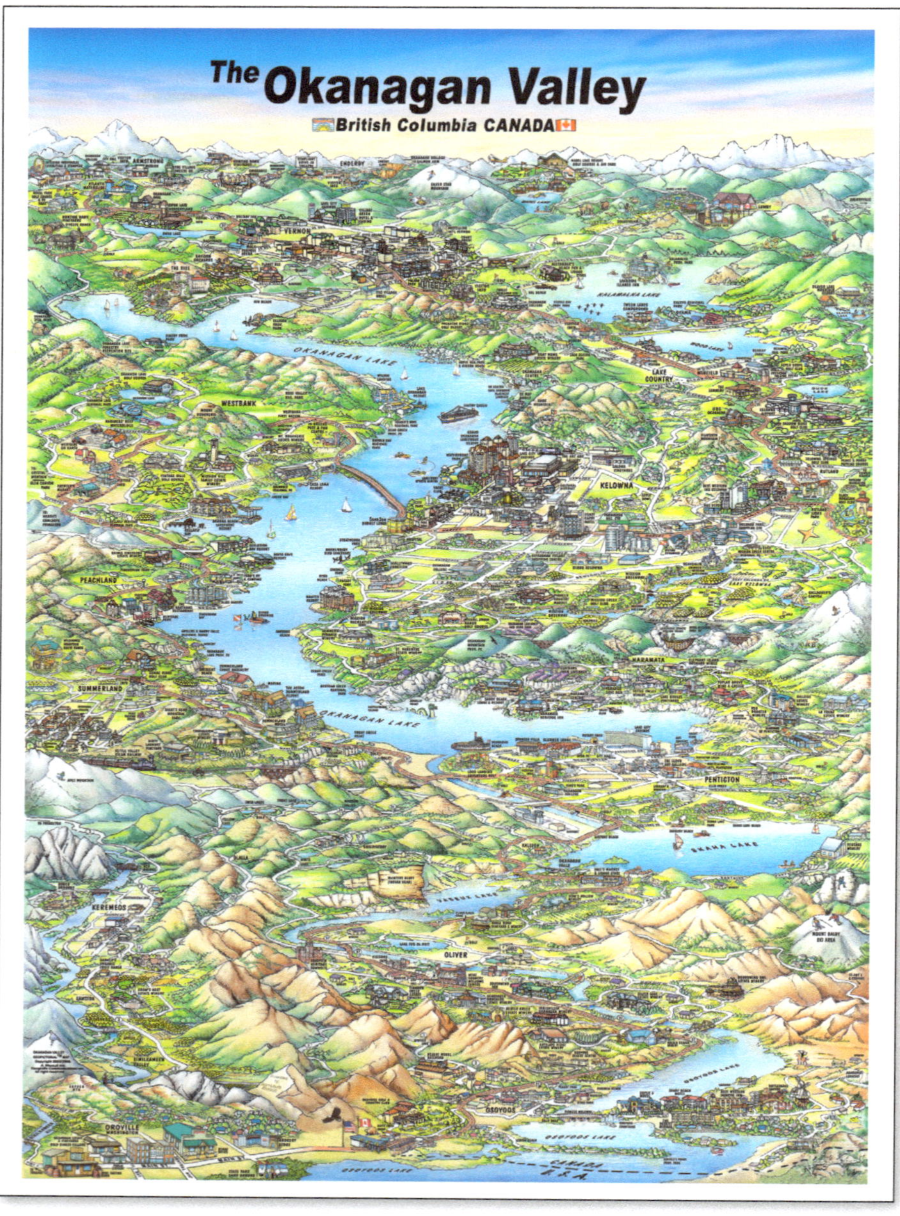

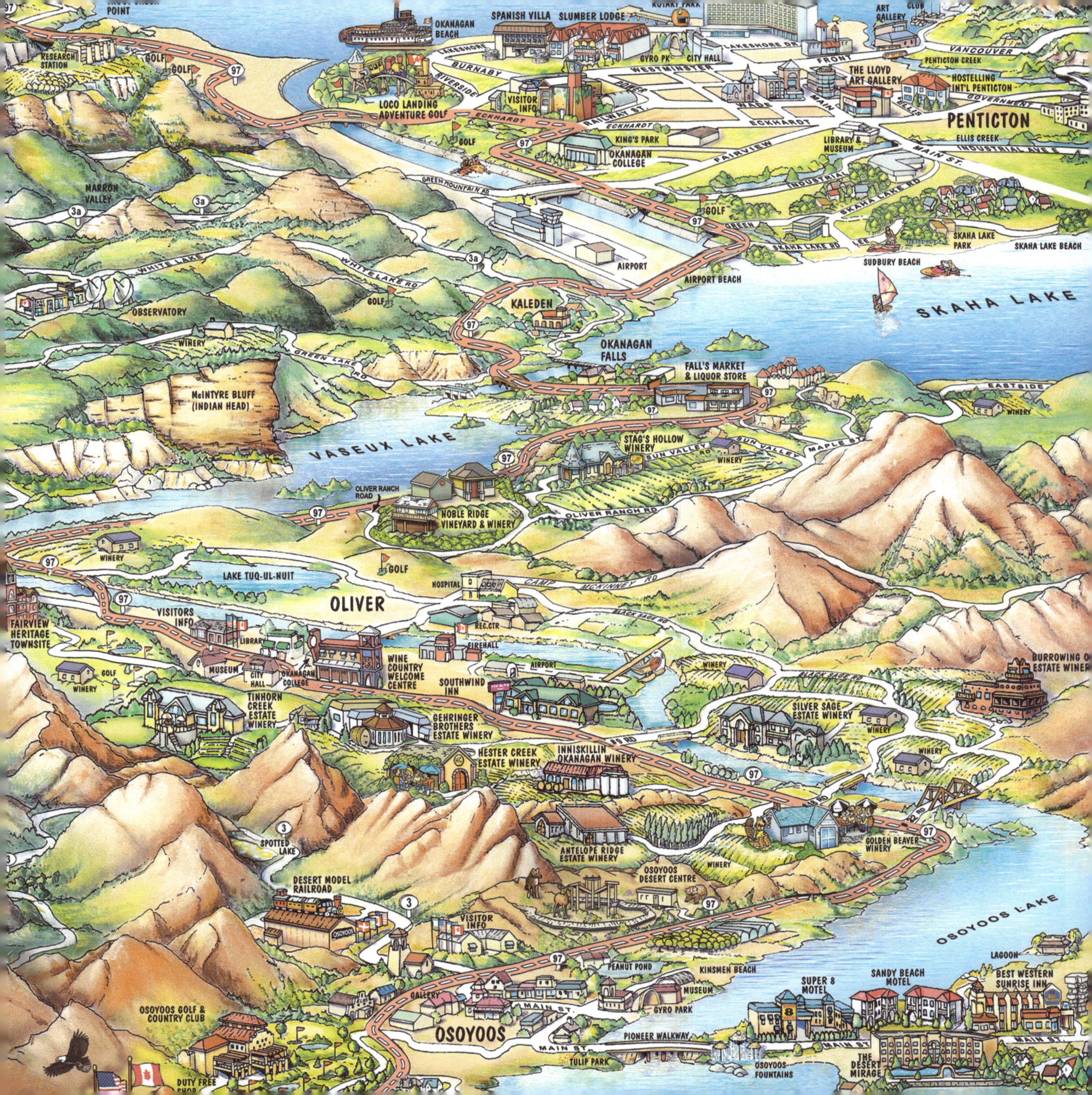

POINT
RESEARCH STATION
GOLF
97
OKANAGAN BEACH
SPANISH VILLA
SLUMBER LODGE
ART GALLERY
GYRO PK
CITY HALL
LAKESHORE DR
VANCOUVER
PENTICTON CREEK
LOCO LANDING ADVENTURE GOLF
RIVERSIDE
BURNABY
WESTMINSTER
FRONT
THE LLOYD ART GALLERY
HOSTELLING INT'L PENTICTON
GOVERNMENT
VISITOR INFO
ECKHARDT
KING'S PARK
OKANAGAN COLLEGE
FAIRVIEW
LIBRARY & MUSEUM
MAIN ST.
PENTICTON
ELLIS CREEK
INDUSTRIAL AVE E
MARRON VALLEY
3a
GREEN MOUNTAIN RD
SKAHA LAKE RD
SKAHA LAKE PARK
SKAHA LAKE BEACH
AIRPORT
AIRPORT BEACH
SUDBURY BEACH
WHITE LAKE
WHITE LAKE RD
OBSERVATORY
KALEDEN
SKAHA LAKE
WINERY
GREEN LAKE RD
OKANAGAN FALLS
FALL'S MARKET & LIQUOR STORE
McINTYRE BLUFF (INDIAN HEAD)
EASTSIDE
VASEUX LAKE
STAG'S HOLLOW WINERY
SUN VALLEY
MAPLE ST
OLIVER RANCH ROAD
NOBLE RIDGE VINEYARD & WINERY
OLIVER RANCH RD
LAKE TUQ-UL-NUIT
HOSPITAL
CAMP MCKINNEY RD
OLIVER
VISITORS INFO
FAIRVIEW HERITAGE TOWNSITE
LIBRARY
REC.CTR
FIREHALL
BLACK SAGE RD
MUSEUM
CITY HALL
OKANAGAN COLLEGE
WINE COUNTRY WELCOME CENTRE
SOUTHWIND INN
AIRPORT
BURROWING OWL ESTATE WINERY
TINHORN CREEK ESTATE WINERY
GEHRINGER BROTHERS ESTATE WINERY
SILVER SAGE ESTATE WINERY
HESTER CREEK ESTATE WINERY
INNISKILLIN OKANAGAN WINERY
SPOTTED LAKE
ANTELOPE RIDGE ESTATE WINERY
GOLDEN BEAVER WINERY
OSOYOOS DESERT CENTRE
DESERT MODEL RAILROAD
VISITOR INFO
OSOYOOS LAKE
PEANUT POND
KINSMEN BEACH
LAGOON
SUPER 8 MOTEL
SANDY BEACH MOTEL
BEST WESTERN SUNRISE INN
GALLERY
MUSEUM
GYRO PARK
MAIN ST.
OSOYOOS GOLF & COUNTRY CLUB
OSOYOOS
PIONEER WALKWAY
TULIP PARK
OSOYOOS FOUNTAINS
THE DESERT MIRAGE
DUTY FREE

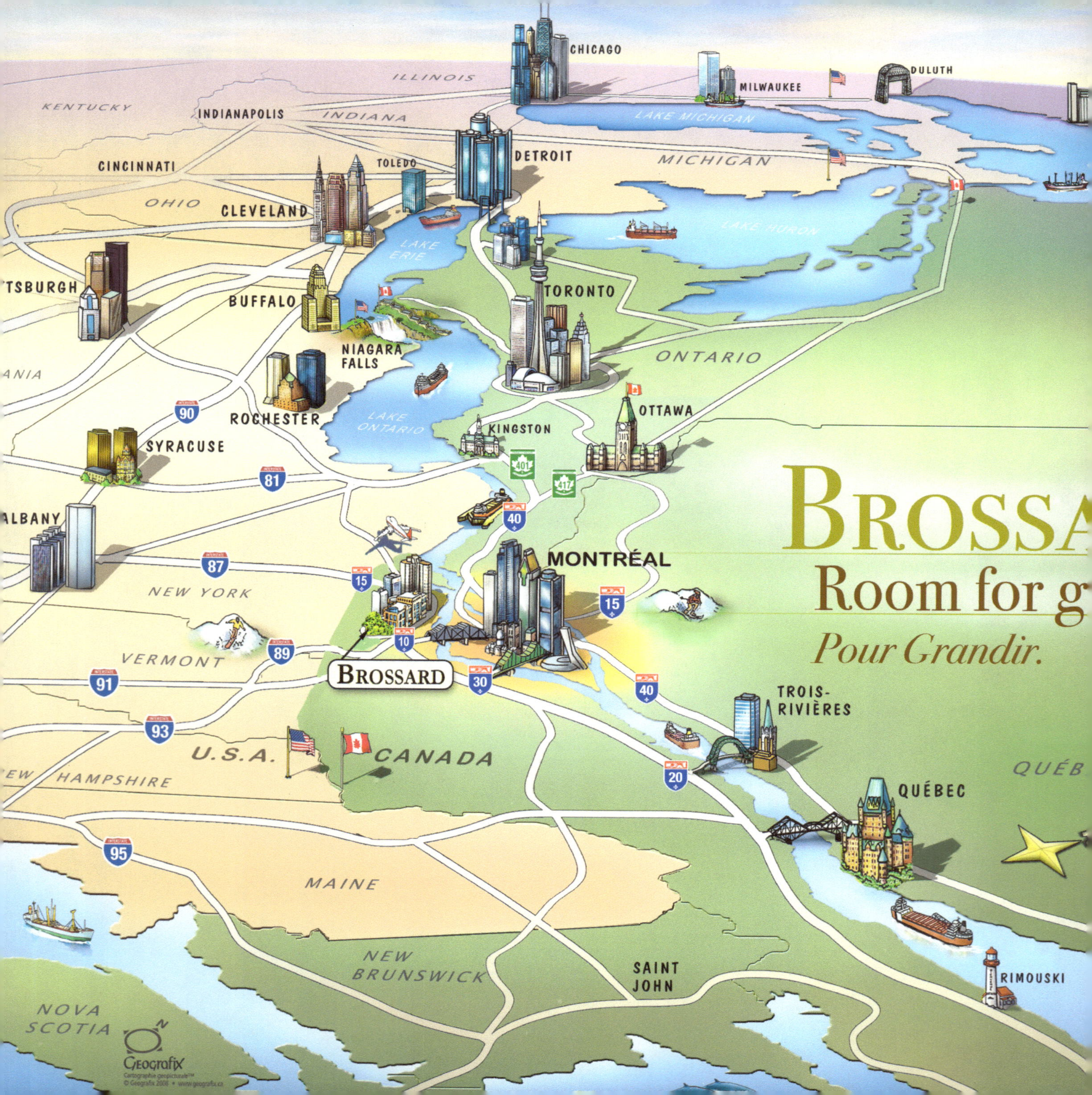

BROSSA
Room for g
Pour Grandir.
CHICAGO
MILWAUKEE
DULUTH
ILLINOIS
KENTUCKY
INDIANAPOLIS
INDIANA
LAKE MICHIGAN
MICHIGAN
CINCINNATI
TOLEDO
DETROIT
OHIO
CLEVELAND
LAKE ERIE
LAKE HURON
TSBURGH
BUFFALO
TORONTO
NIAGARA FALLS
ONTARIO
ANIA
90
ROCHESTER
LAKE ONTARIO
KINGSTON
OTTAWA
SYRACUSE
401
417
81
ALBANY
40
MONTRÉAL
87
15
NEW YORK
15
VERMONT
89
10
BROSSARD
91
30
40
TROIS-RIVIÈRES
93
U.S.A.
CANADA
20
EW HAMPSHIRE
QUÉB
QUÉBEC
95
MAINE
NEW BRUNSWICK
SAINT JOHN
RIMOUSKI
NOVA SCOTIA
N
GEOGRAFIX

Jean-Louis Rheault, Geografix

Brossard (North America) Geopictorial™ map

Media used: Ink, Watercolour, AdobePhotoshop

More information: www.geografix.ca

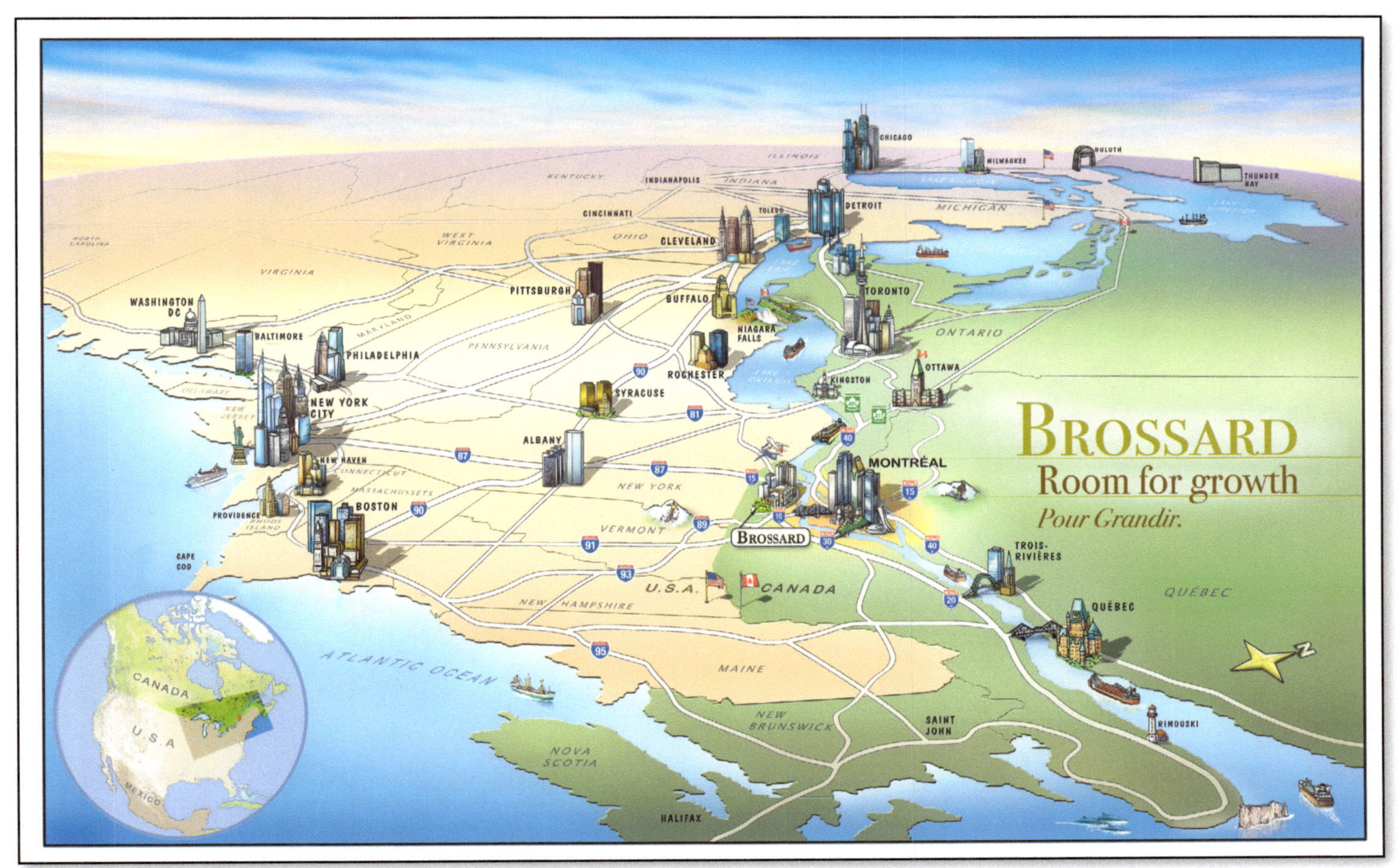

Dale Sanderson, Dex (R.H. Donnelley)

Street map of the Grand Valley

Colorado, USA

Software used: ESRI ArcGIS, Macromedia FreeHand MX, Natural Scene Designer, MacDEM, Adobe Photoshop.

Data sources: Most planimetric data from Mesa County GIS. Some data was obtained and/or updated based on an on-site field-check. DEMs from USGS.

More information: www.usends.com/mapguy/Portfolio/

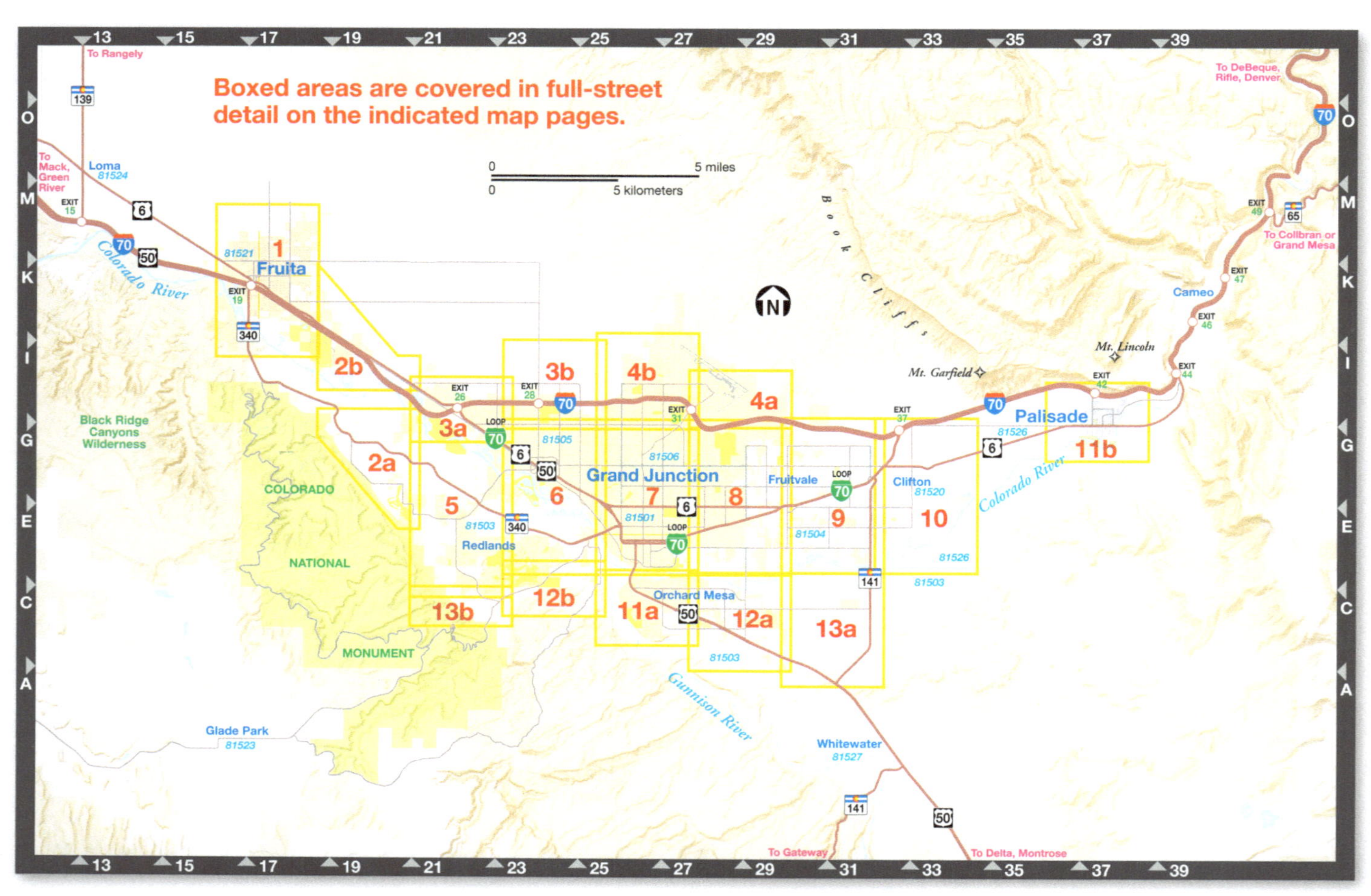

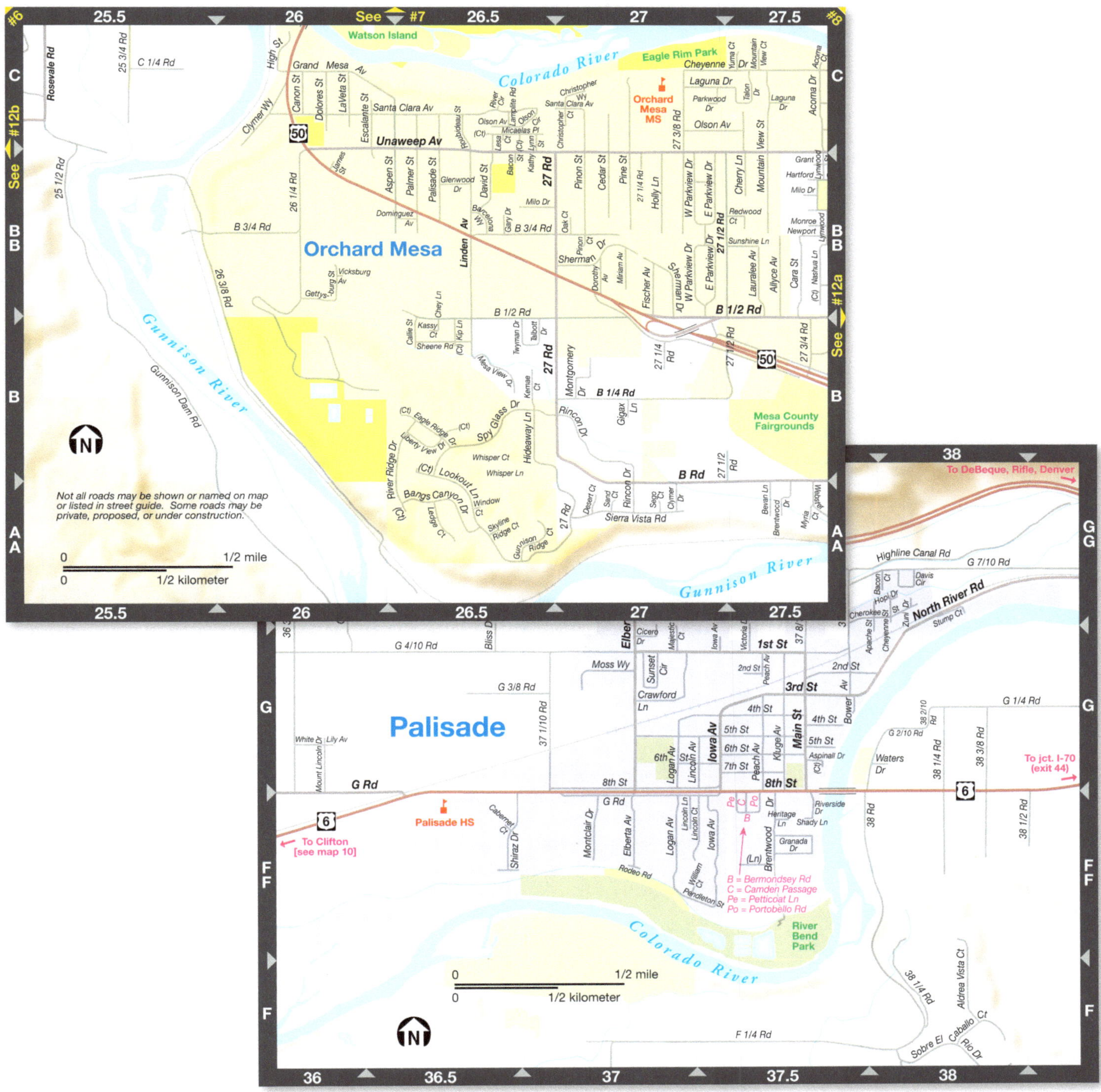

Orchard Mesa
Watson Island
Eagle Rim Park
Colorado River
Gunnison River
Orchard Mesa MS
Mesa County Fairgrounds
Unaweep Av
Santa Clara Av
Grand Mesa Av
B 3/4 Rd
B 1/2 Rd
B 1/4 Rd
B Rd
C 1/4 Rd
27 Rd
Linden Av
Gunnison Dam Rd
Sierra Vista Rd
Not all roads may be shown or named on map or listed in street guide. Some roads may be private, proposed, or under construction.
0 1/2 mile
0 1/2 kilometer
See #7
See #12b
See #12a
#6
#8
Palisade
Palisade HS
River Bend Park
G Rd
G 4/10 Rd
G 3/8 Rd
G 1/4 Rd
F 1/4 Rd
Highline Canal Rd
North River Rd
Main St
Iowa Av
3rd St
8th St
To DeBeque, Rifle, Denver
To jct. I-70 (exit 44)
To Clifton [see map 10]
B = Bermondsey Rd
C = Camden Passage
Pe = Petticoat Ln
Po = Portobello Rd
0 1/2 mile
0 1/2 kilometer

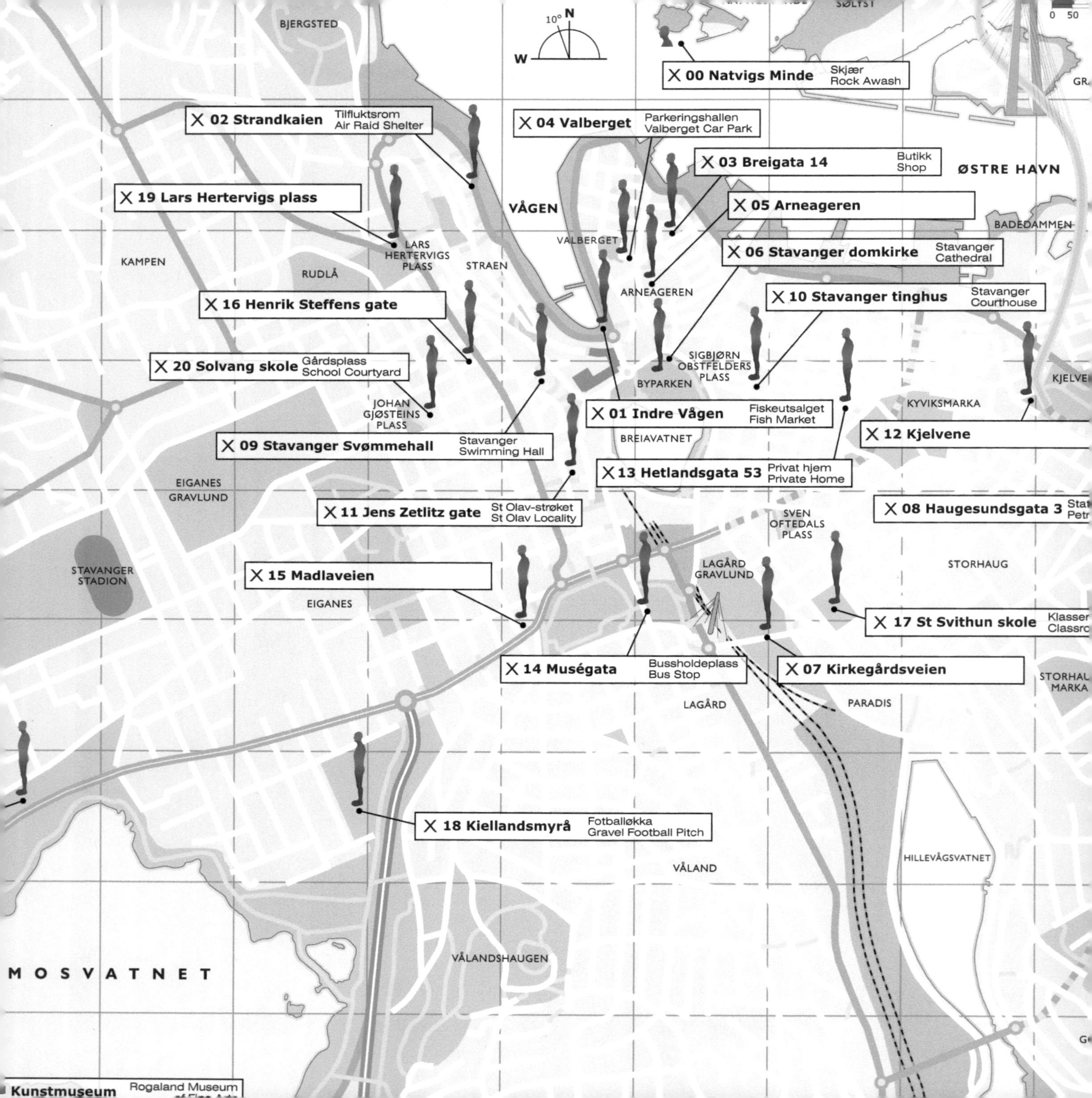

X 00 Natvigs Minde Skjær Rock Awash
X 01 Indre Vågen Fiskeutsalget Fish Market
X 02 Strandkaien Tilfluktsrom Air Raid Shelter
X 03 Breigata 14 Butikk Shop
X 04 Valberget Parkeringshallen Valberget Car Park
X 05 Arneageren
X 06 Stavanger domkirke Stavanger Cathedral
X 07 Kirkegårdsveien
X 08 Haugesundsgata 3
X 09 Stavanger Svømmehall Stavanger Swimming Hall
X 10 Stavanger tinghus Stavanger Courthouse
X 11 Jens Zetlitz gate St Olav-strøket St Olav Locality
X 12 Kjelvene
X 13 Hetlandsgata 53 Privat hjem Private Home
X 14 Muségata Bussholdeplass Bus Stop
X 15 Madlaveien
X 16 Henrik Steffens gate
X 17 St Svithun skole Klasser
X 18 Kiellandsmyrå Fotballøkka Gravel Football Pitch
X 19 Lars Hertervigs plass
X 20 Solvang skole Gårdsplass School Courtyard
Kunstmuseum Rogaland Museum
N
W
10°
0 50
BJERGSTED
SØLYST
VÅGEN
ØSTRE HAVN
BADEDAMMEN
KAMPEN
RUDLÅ
LARS HERTERVIGS PLASS
STRAEN
VALBERGET
ARNEAGEREN
SIGBJØRN OBSTFELDERS PLASS
BYPARKEN
KYVIKSMARKA
JOHAN GJØSTEINS PLASS
BREIAVATNET
EIGANES GRAVLUND
SVEN OFTEDALS PLASS
STORHAUG
STAVANGER STADION
LAGÅRD GRAVLUND
EIGANES
LAGÅRD
PARADIS
STORHAUG MARKA
HILLEVÅGSVATNET
VÅLAND
VÅLANDSHAUGEN
M O S V A T N E T

Kevin Paul Scarrott, Stavanger Guide Maps Norway

Antony Gormley - Broken Column, Norway

Software used: Adobe Illustrator, CorelDraw, Adobe Photoshop

Data sources: Stavanger Municipality, Norway

More information: www.stavanger-guide.no

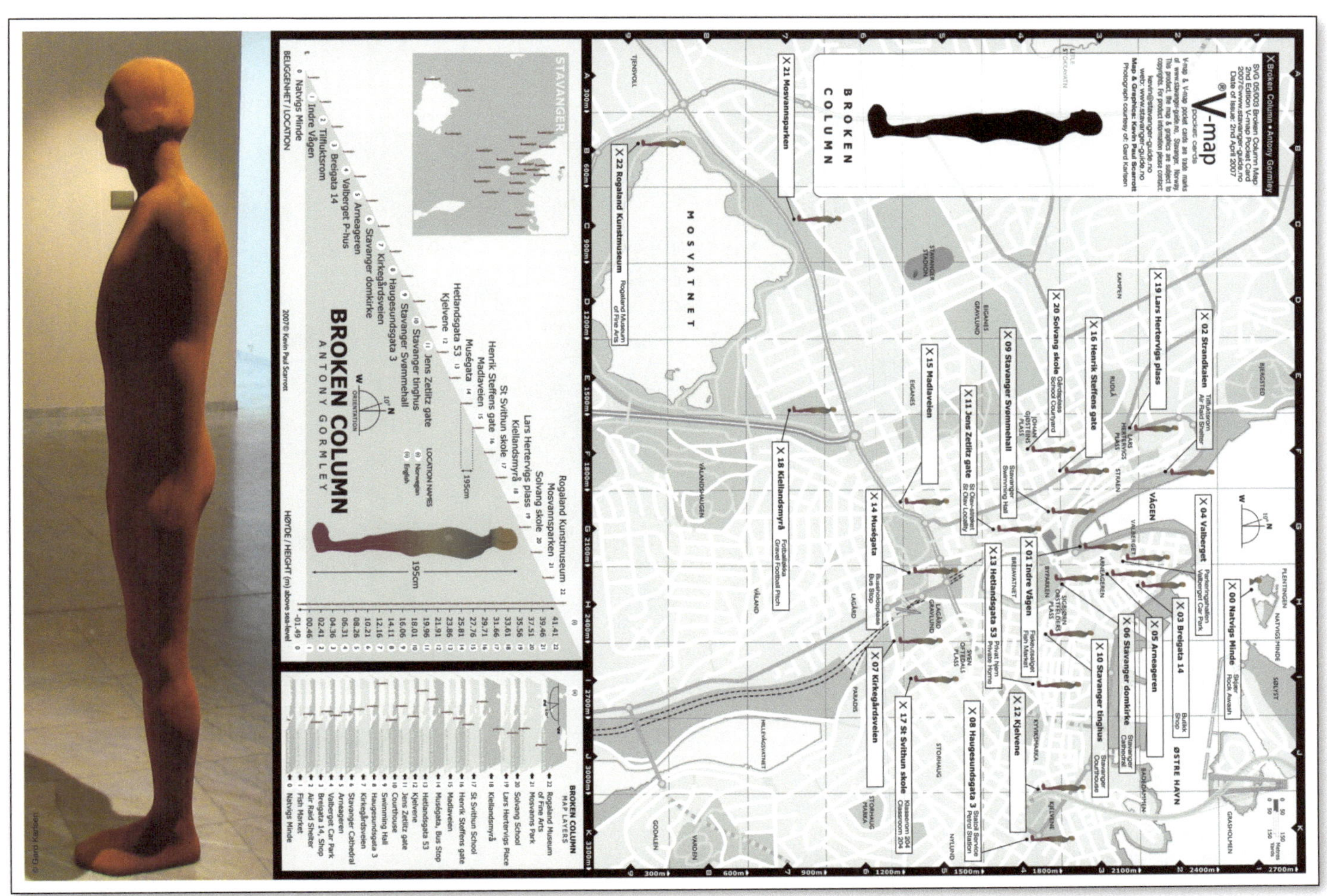

Michael Scisco, BioGeoCreations

Inland Northwest Land Trust Service Area

Software used: ESRI ArcView, Adobe Photoshop, Adobe Illustrator

More information: www.biogeocreations.com

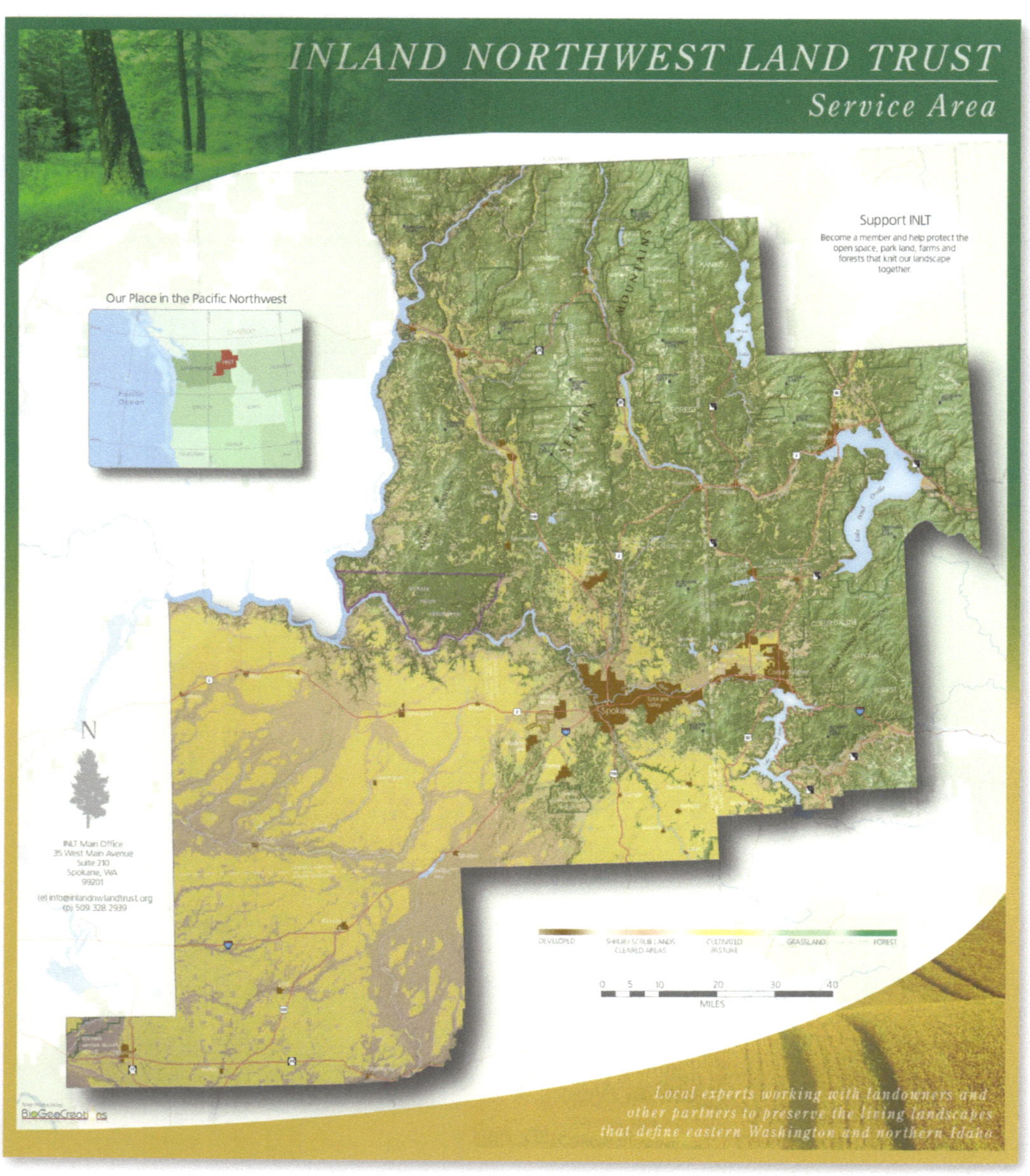

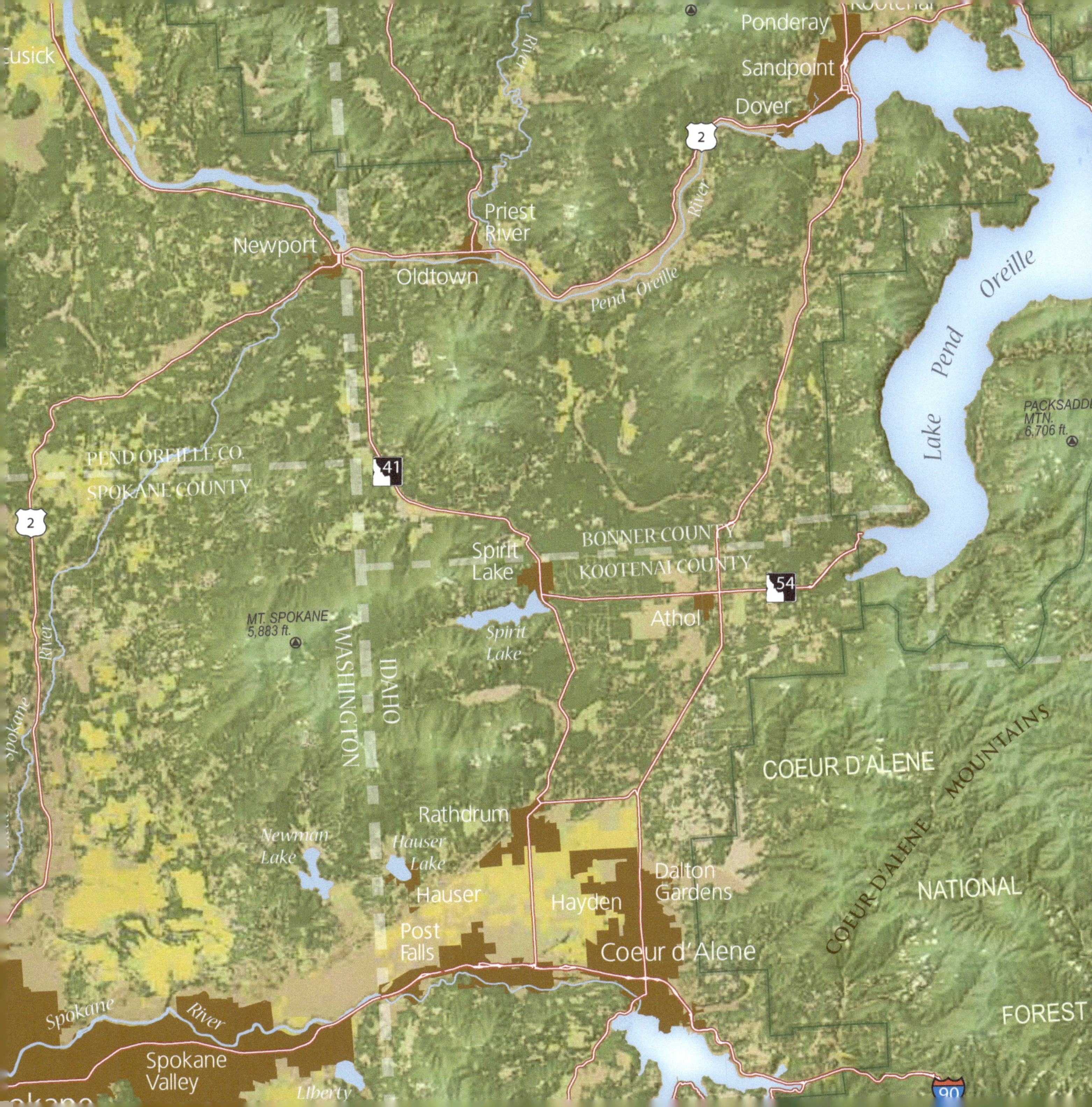

Cusick
Ponderay
Kootenai
Sandpoint
Dover
2
River
Priest River
River
Newport
Oldtown
Pend Oreille
Lake Pend Oreille
PACKSADDLE MTN. 6,706 ft.
PEND OREILLE CO.
SPOKANE COUNTY
41
2
Spirit Lake
BONNER COUNTY
KOOTENAI COUNTY
54
Athol
MT. SPOKANE 5,883 ft.
Spirit Lake
WASHINGTON
IDAHO
Spokane River
COEUR D'ALENE MOUNTAINS
Rathdrum
Newman Lake
Hauser Lake
Hauser
Hayden
Dalton Gardens
COEUR D'ALENE NATIONAL FOREST
Post Falls
Coeur d'Alene
Spokane
River
Spokane Valley
Liberty
90

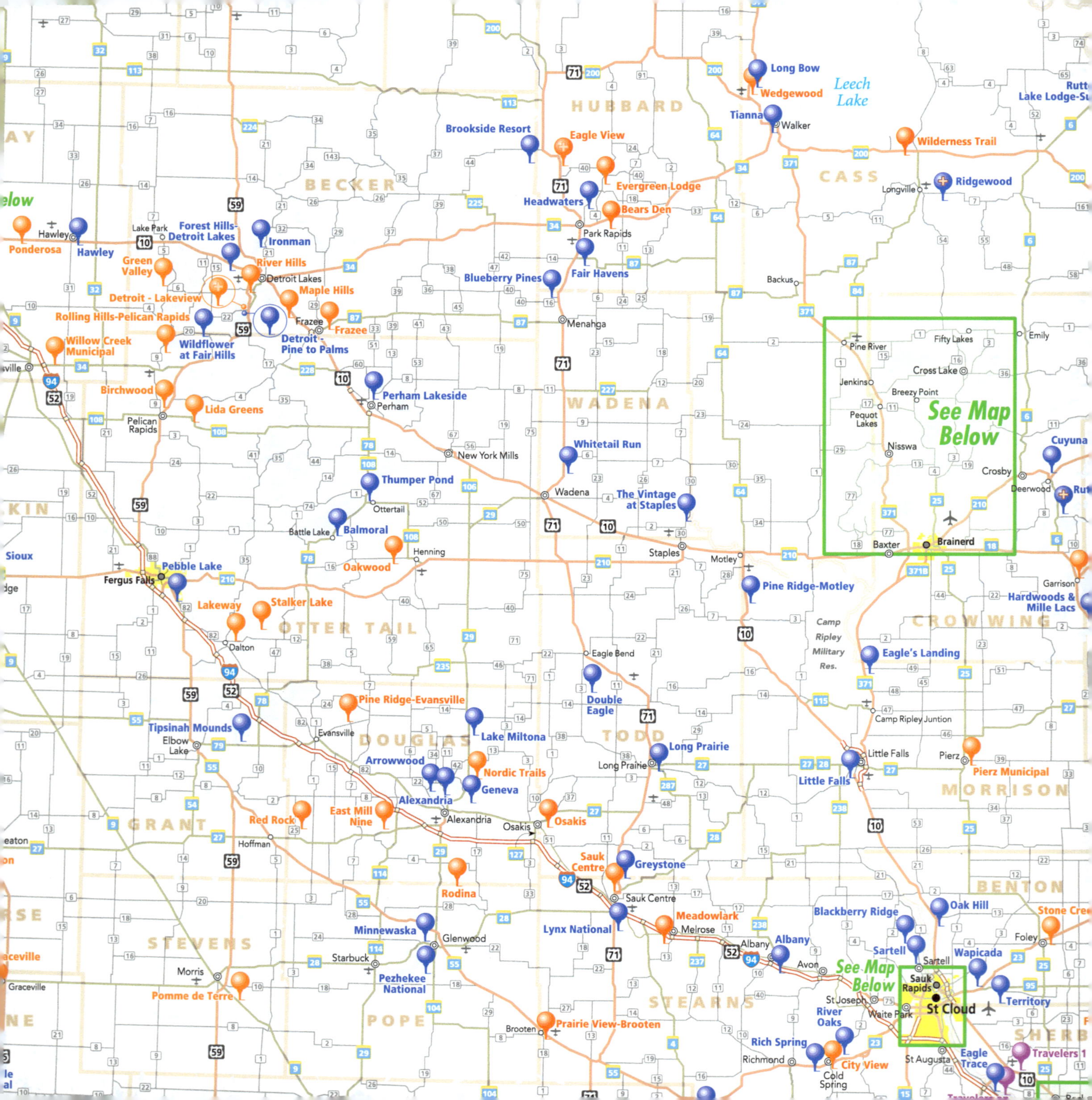

Long Bow
Wedgewood
Leech Lake
Tianna
Walker
Rutt
Lake Lodge-Su
HUBBARD
Eagle View
Brookside Resort
Wilderness Trail
CASS
Longville
Ridgewood
BECKER
Evergreen Lodge
Headwaters
Bears Den
Park Rapids
Ponderosa
Hawley
Hawley
Lake Park
Forest Hills-Detroit Lakes
Ironman
Green Valley
River Hills
Detroit Lakes
Fair Havens
Blueberry Pines
Backus
Detroit - Lakeview
Maple Hills
Rolling Hills-Pelican Rapids
Frazee
Frazee
Detroit - Pine to Palms
Menahga
Willow Creek Municipal
Wildflower at Fair Hills
Pine River
Fifty Lakes
Emily
Cross Lake
Jenkins
Breezy Point
Pequot Lakes
See Map Below
Nisswa
Crosby
Cuyuna
Deerwood
Ruttger
Birchwood
Perham Lakeside
Perham
Lida Greens
Pelican Rapids
WADENA
New York Mills
Whitetail Run
Thumper Pond
Ottertail
Wadena
The Vintage at Staples
Balmoral
Battle Lake
Henning
Staples
Motley
Baxter
Brainerd
Sioux
Pebble Lake
Fergus Falls
Oakwood
Pine Ridge-Motley
Garrison
Hardwoods & Mille Lacs
Lakeway
Stalker Lake
OTTER TAIL
Camp Ripley Military Res.
CROW WING
Dalton
Eagle Bend
Eagle's Landing
Double Eagle
Pine Ridge-Evansville
Camp Ripley Juntion
Tipsinah Mounds
Elbow Lake
Evansville
Lake Miltona
DOUGLAS
TODD
Long Prairie
Long Prairie
Little Falls
Pierz
Arrowwood
Nordic Trails
Geneva
Little Falls
Pierz Municipal
Alexandria
MORRISON
Red Rock
East Mill Nine
Alexandria
Osakis
Osakis
GRANT
Hoffman
Sauk Centre
Greystone
Rodina
Sauk Centre
BENTON
Oak Hill
Stone Creek
Minnewaska
Meadowlark
Lynx National
Melrose
Blackberry Ridge
Foley
Glenwood
Albany
Albany
Sartell
Wapicada
STEVENS
Starbuck
Avon
Sartell
See Map Below
Sauk Rapids
Morris
Pezhekee National
St Joseph
St Cloud
Territory
Graceville
Pomme de Terre
POPE
STEARNS
River Oaks
Waite Park
Prairie View-Brooten
Brooten
Rich Spring
Eagle Trace
Travelers
Richmond
St Augusta
City View
Cold Spring
SHERB

Nate Sievers, Nat Case, Hedberg Maps, Inc.

Minnesota Golf Map

Software used: Macromedia Freehand, Adobe Illustrator, Avenza MAPublisher, Adobe InDesign

More information: www.hedbergmaps.com

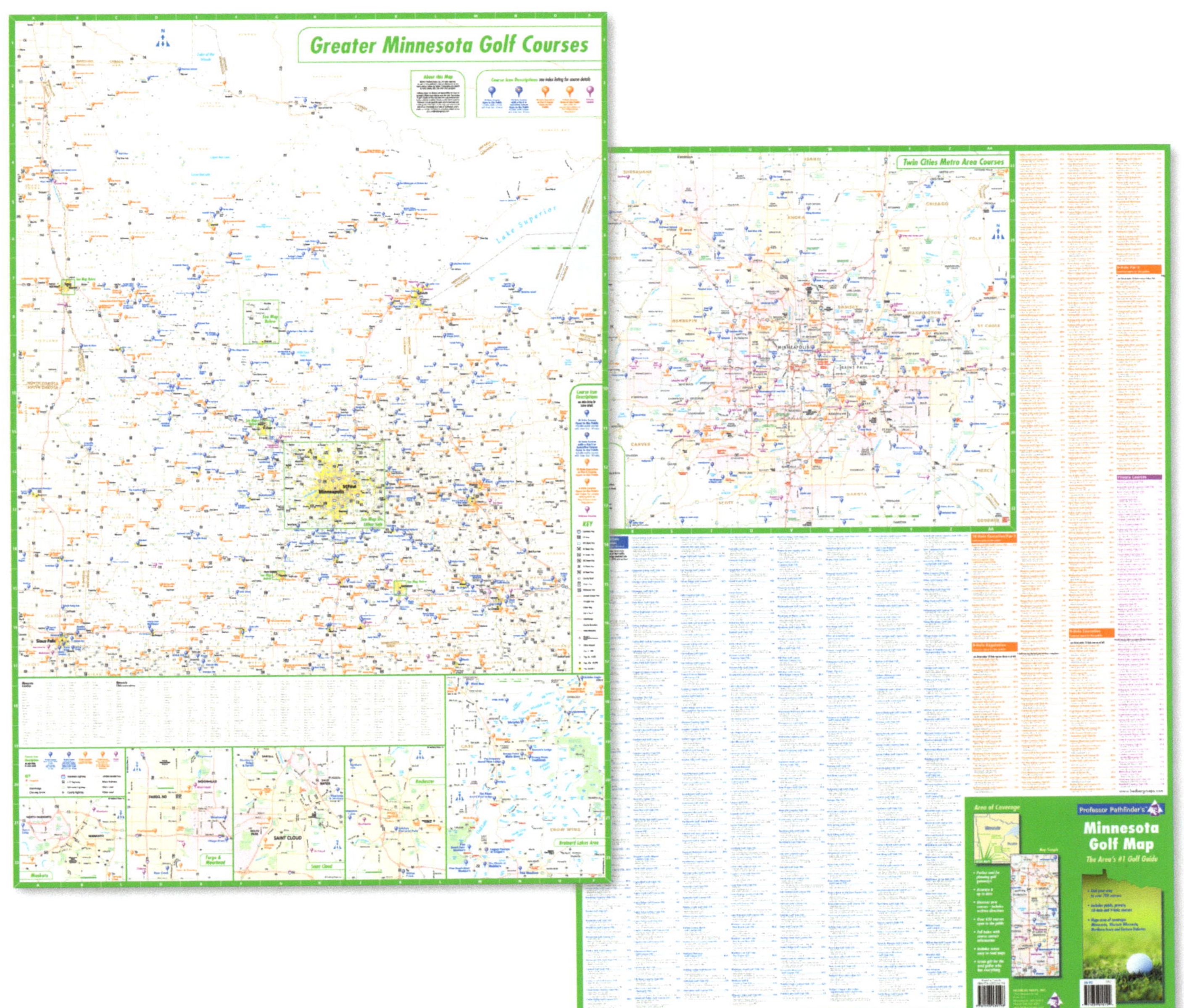

Nick Springer, Springer Cartographics LLC

Colorado Ground Snow Loads

www.seacolorado.org

Software used: Manifold System, Surfer, Adobe Illustrator, Adobe Photoshop

More information: www.springercartographics.com

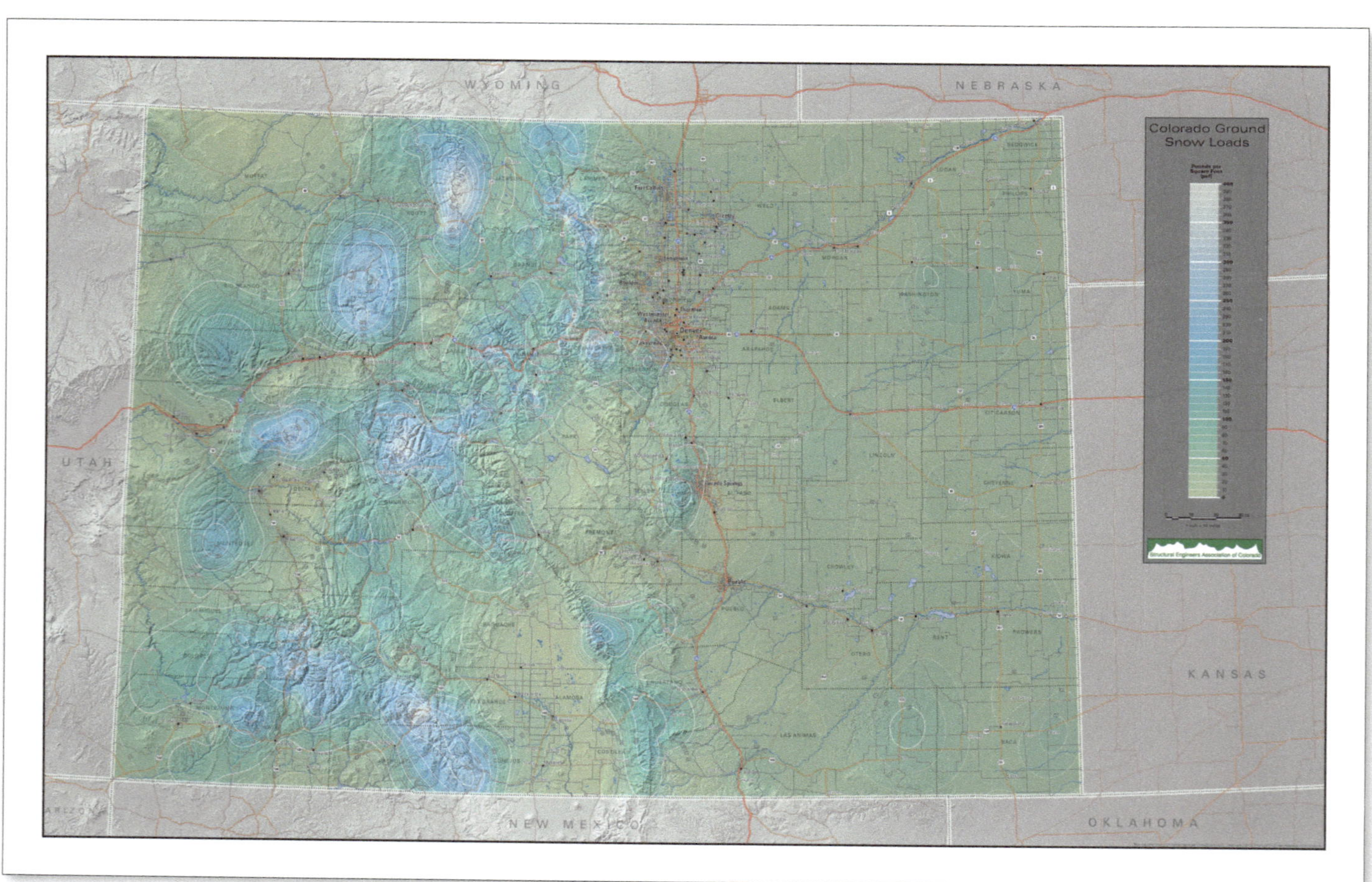

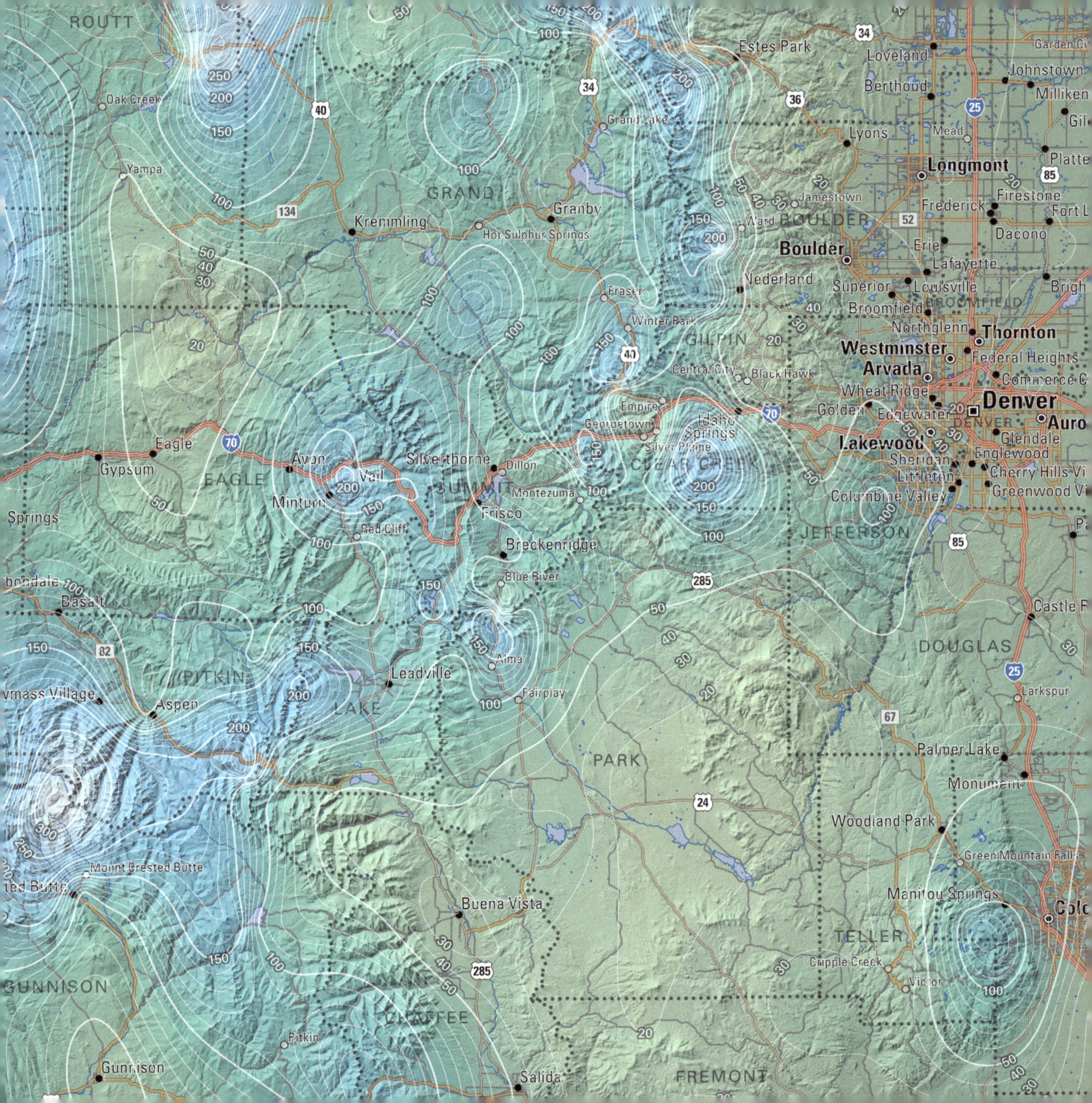

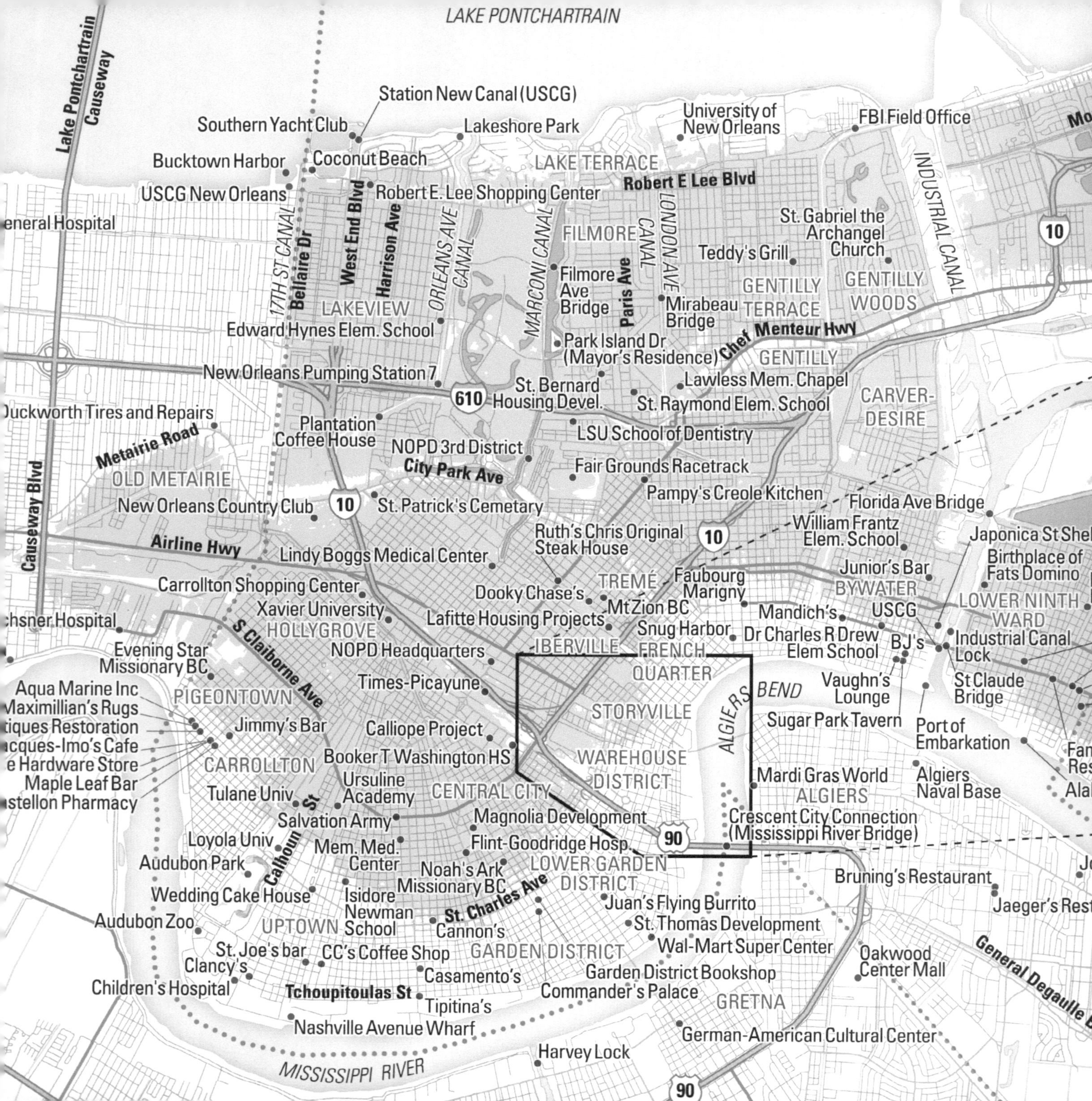
LAKE PONTCHARTRAIN
Lake Pontchartrain Causeway
Station New Canal (USCG)
Southern Yacht Club
Lakeshore Park
University of New Orleans
FBI Field Office
Bucktown Harbor
Coconut Beach
LAKE TERRACE
USCG New Orleans
Robert E. Lee Shopping Center
Robert E Lee Blvd
eneral Hospital
17TH ST CANAL
Bellaire Dr
West End Blvd
Harrison Ave
ORLEANS AVE CANAL
MARCONI CANAL
FILMORE
LONDON AVE CANAL
INDUSTRIAL CANAL
St. Gabriel the Archangel Church
Teddy's Grill
10
Filmore Ave Bridge
Paris Ave
GENTILLY TERRACE
GENTILLY WOODS
LAKEVIEW
Mirabeau Bridge
Edward Hynes Elem. School
Park Island Dr (Mayor's Residence)
Chef Menteur Hwy
GENTILLY
New Orleans Pumping Station 7
St. Bernard Housing Devel.
Lawless Mem. Chapel
610
St. Raymond Elem. School
CARVER-DESIRE
Duckworth Tires and Repairs
Plantation Coffee House
LSU School of Dentistry
Metairie Road
NOPD 3rd District
City Park Ave
Fair Grounds Racetrack
OLD METAIRIE
Pampy's Creole Kitchen
Causeway Blvd
New Orleans Country Club
St. Patrick's Cemetary
Florida Ave Bridge
Ruth's Chris Original Steak House
William Frantz Elem. School
Japonica St Shel
Airline Hwy
Lindy Boggs Medical Center
Birthplace of Fats Domino
Junior's Bar
Carrollton Shopping Center
TREMÉ
Faubourg Marigny
BYWATER
Dooky Chase's
LOWER NINTH WARD
Xavier University
Mt Zion BC
Mandich's
USCG
chsner Hospital
Lafitte Housing Projects
HOLLYGROVE
Snug Harbor
Dr Charles R Drew Elem School
BJ's
Industrial Canal Lock
S Claiborne Ave
Evening Star Missionary BC
NOPD Headquarters
IBERVILLE
FRENCH QUARTER
Aqua Marine Inc
Maximillian's Rugs
Times-Picayune
Vaughn's Lounge
St Claude Bridge
PIGEONTOWN
ALGIERS BEND
STORYVILLE
tiques Restoration
Jimmy's Bar
Calliope Project
Sugar Park Tavern
acques-Imo's Cafe
Port of Embarkation
e Hardware Store
CARROLLTON
Booker T Washington HS
WAREHOUSE DISTRICT
Fam Res
Maple Leaf Bar
Mardi Gras World
Algiers Naval Base
astellon Pharmacy
Tulane Univ
Ursuline Academy
CENTRAL CITY
ALGIERS
Ala
Salvation Army
Magnolia Development
Crescent City Connection (Mississippi River Bridge)
Loyola Univ
Calhoun St
Mem. Med. Center
Flint-Goodridge Hosp.
90
Audubon Park
Noah's Ark Missionary BC
LOWER GARDEN DISTRICT
Bruning's Restaurant
Wedding Cake House
Isidore Newman School
St. Charles Ave
Juan's Flying Burrito
Jaeger's Rest
Audubon Zoo
UPTOWN
Cannon's
St. Thomas Development
St. Joe's bar
CC's Coffee Shop
GARDEN DISTRICT
Wal-Mart Super Center
Oakwood Center Mall
Clancy's
Casamento's
Garden District Bookshop
General Degaulle B
Children's Hospital
Tchoupitoulas St
Commander's Palace
GRETNA
Tipitina's
Nashville Avenue Wharf
German-American Cultural Center
Harvey Lock
MISSISSIPPI RIVER
90

Nick Springer, Springer Cartographics LLC

New Orleans and Gulf Coast
From "The Great Deluge" by Douglas Brinkley

Software used: Manifold System, Adobe Illustrator
More information: www.springercartographics.com

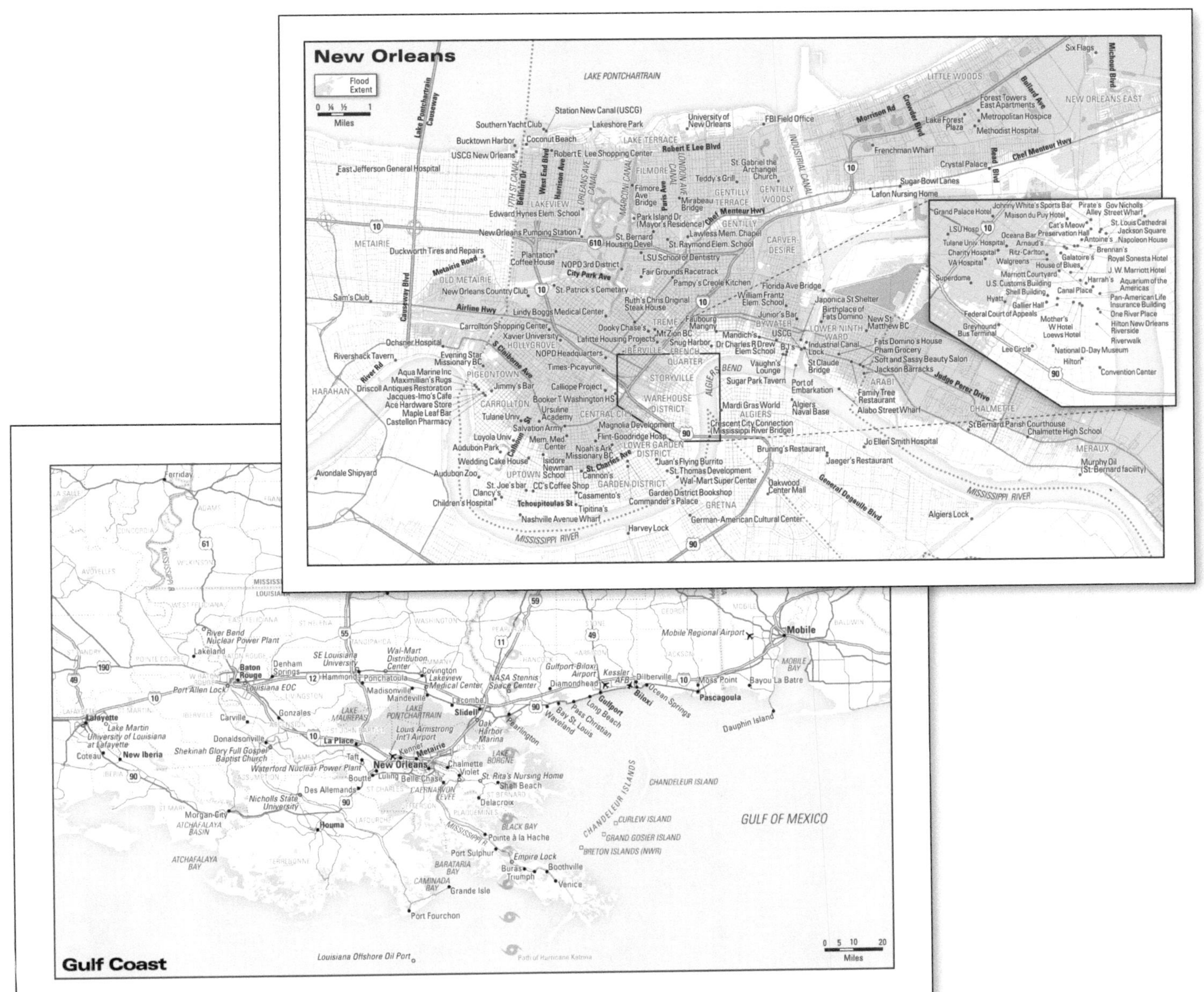

Charles Syrett, Map Graphics

Missoula Key Map

Software used: Adobe FreeHand, Avenza MAPublisher, GeoCart, Adobe Photoshop

Data sources: United States Geological Survey

More information: www.mapgraphics.com

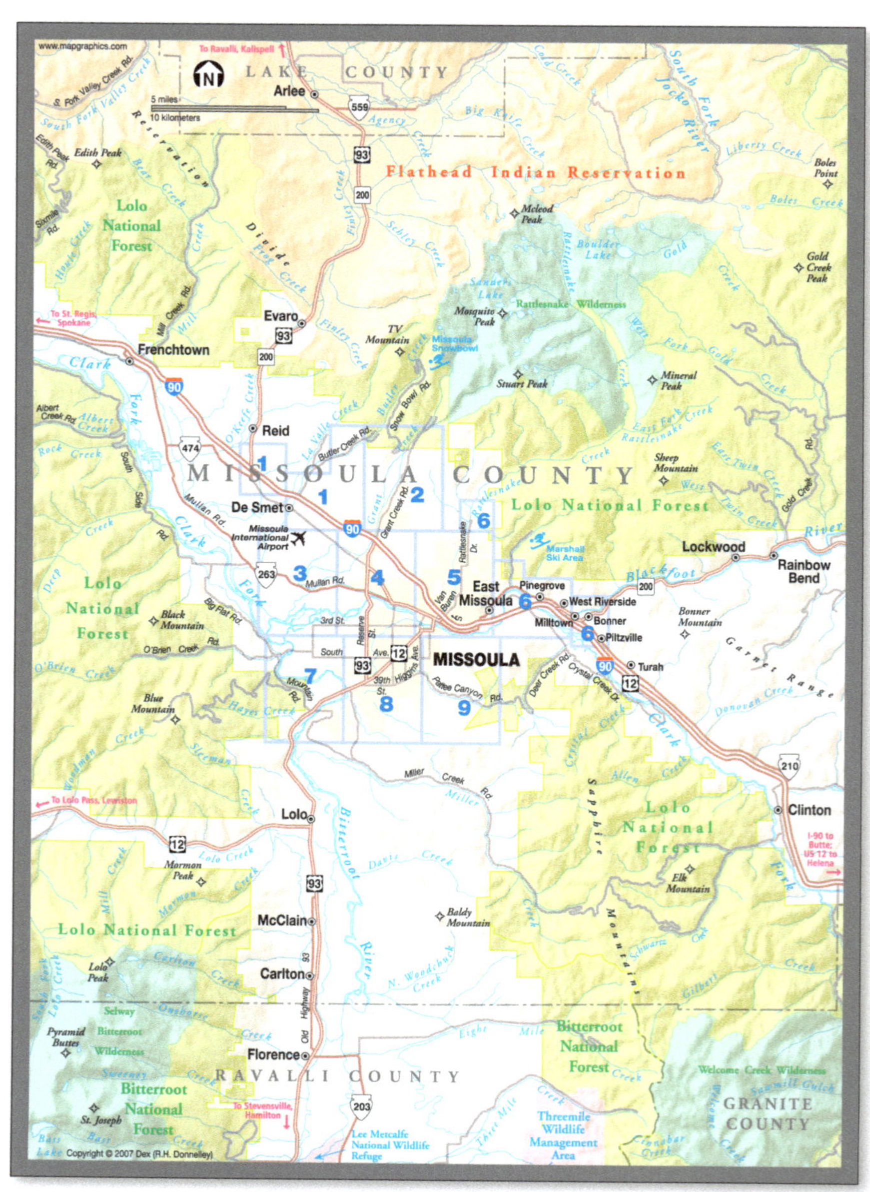

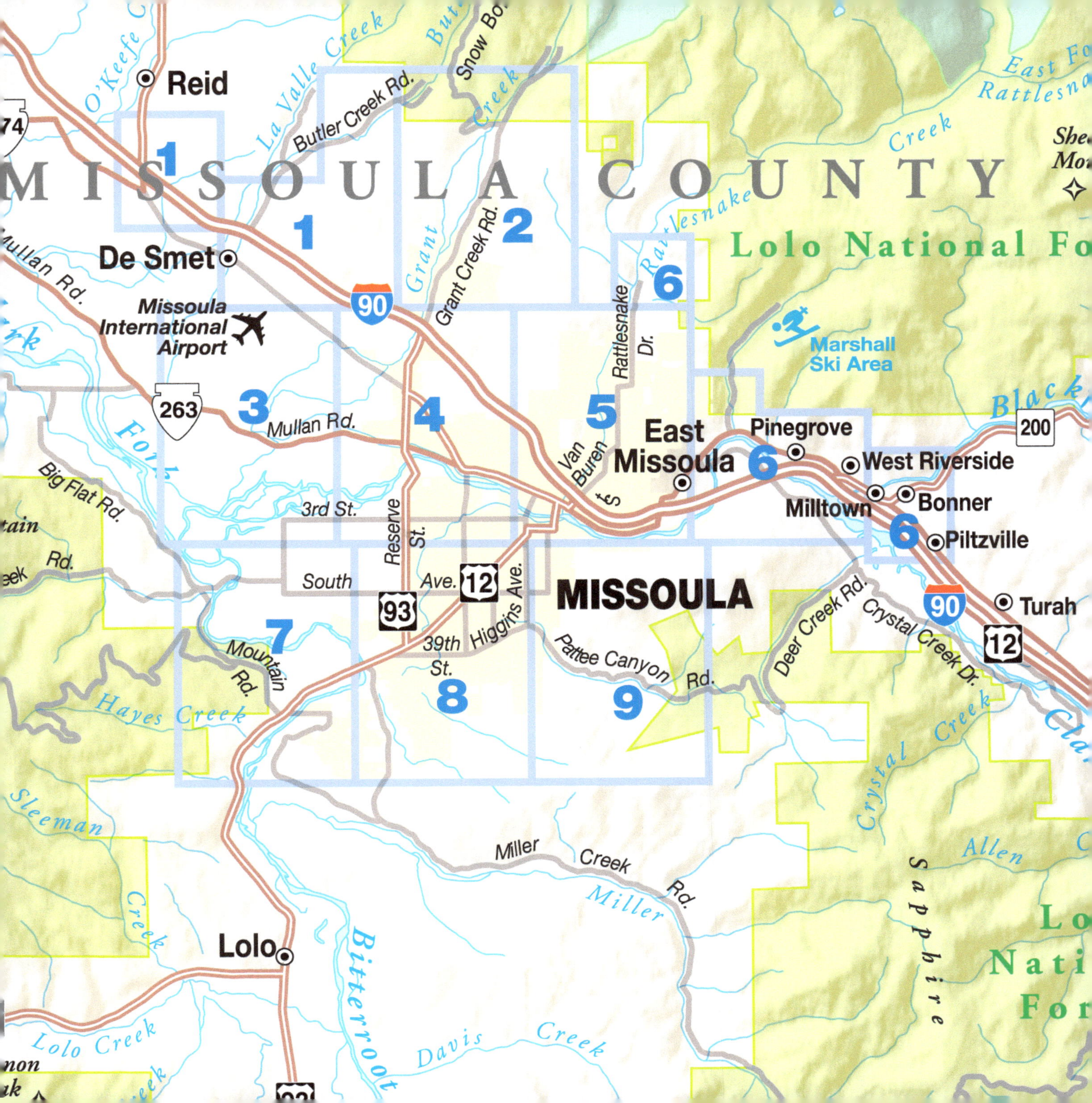
MISSOULA COUNTY
Lolo National Forest
Reid
De Smet
Missoula International Airport
East Missoula
Pinegrove
West Riverside
Milltown
Bonner
Piltzville
Turah
MISSOULA
Lolo
Marshall Ski Area
Butler Creek Rd.
Grant Creek Rd.
Rattlesnake Dr.
Mullan Rd.
Big Flat Rd.
3rd St.
Reserve St.
South Ave.
Higgins Ave.
39th St.
Mountain Rd.
Pattee Canyon Rd.
Deer Creek Rd.
Crystal Creek Dr.
Miller Creek Rd.
Van Buren St.
O'Keefe Creek
La Valle Creek
Grant Creek
Snow Bowl
Rattlesnake Creek
East Fork Rattlesnake Creek
Hayes Creek
Sleeman Creek
Lolo Creek
Bitterroot
Miller Creek
Davis Creek
Crystal Creek
Allen Creek
Sapphire
90
263
200
12
93
1
2
3
4
5
6
7
8
9

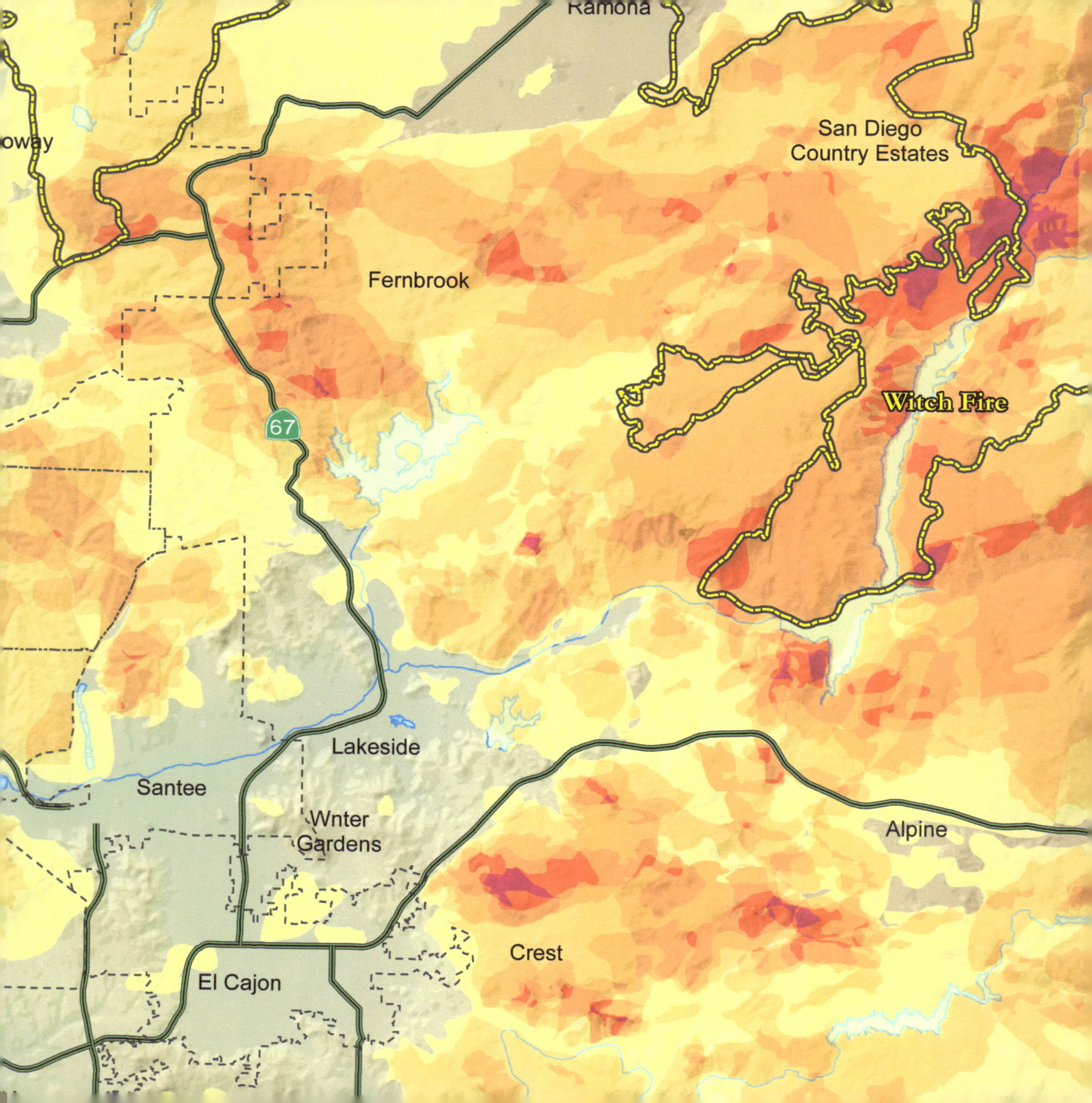
Ramona
oway
San Diego
Country Estates
Fernbrook
67
Witch Fire
Lakeside
Santee
Wnter
Gardens
Alpine
Crest
El Cajon

David Toney, GISP, Flat Planet Maps

San Diego County Fire History Count 1910-2006 with 2007 Fires

Software used: ESRI ArcGIS

Data sources: San Diego County Association of Governments (SANDAG), San Diego Geographic Information Source (SanGIS), California Department of Forestry and Fire Protection

More information: www.flatplanetmaps.com

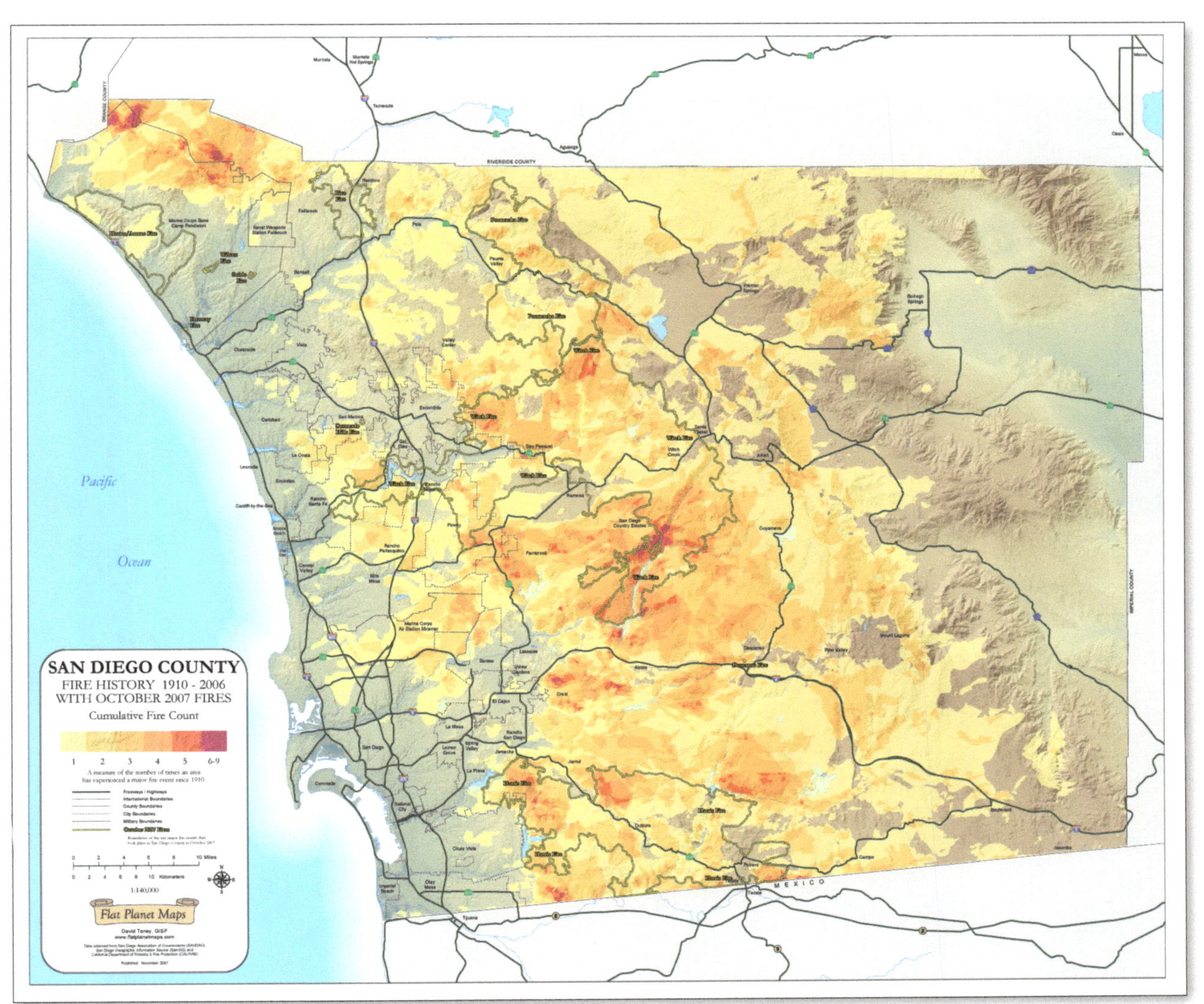

Hans van der Maarel, Red Geographics

2008 Apeldoorn City Map

Copyright © 2007 City of Apeldoorn

Software used: FME, Adobe Illustrator, Avenza MAPublisher

More information: www.redgeographics.com

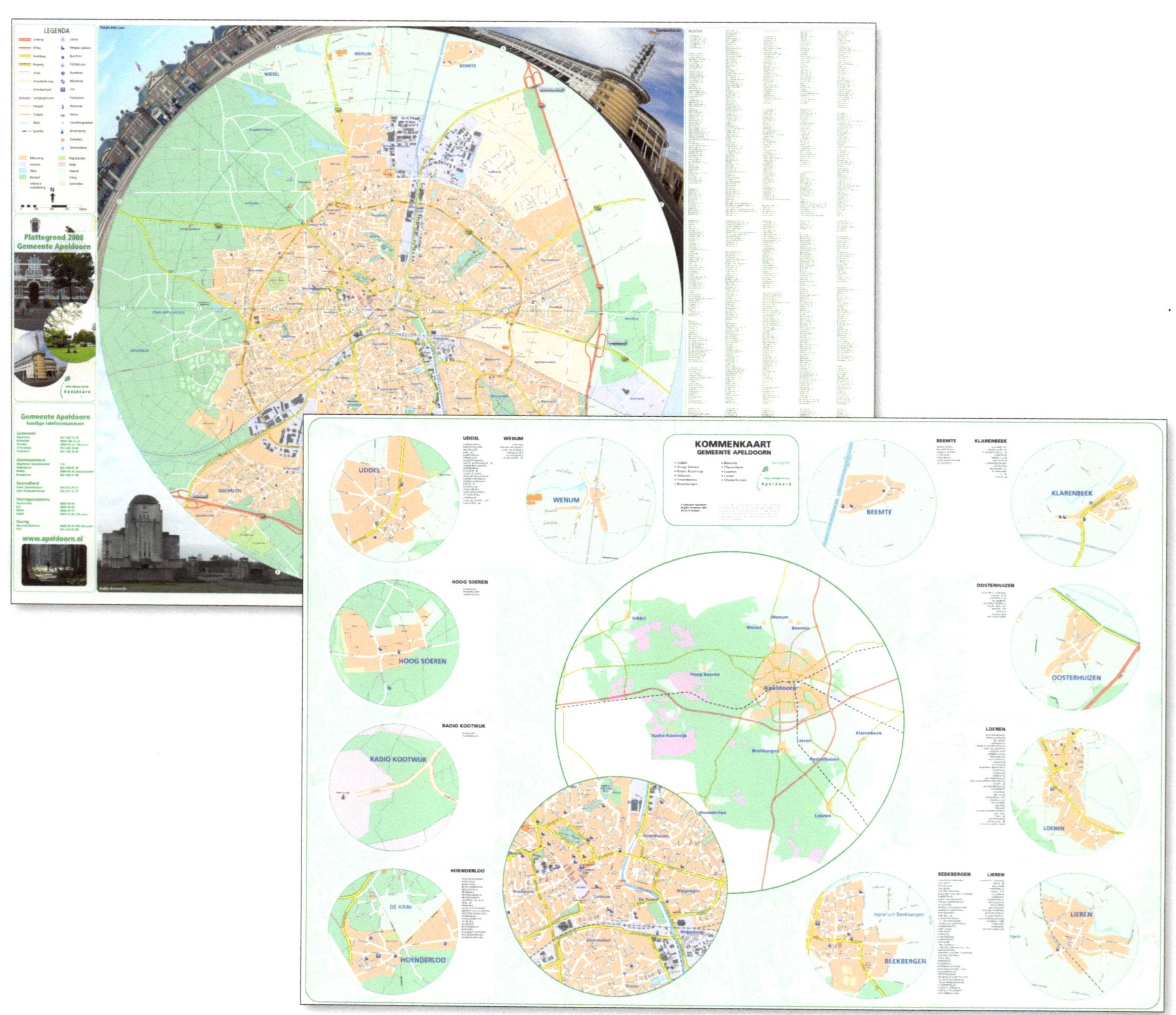

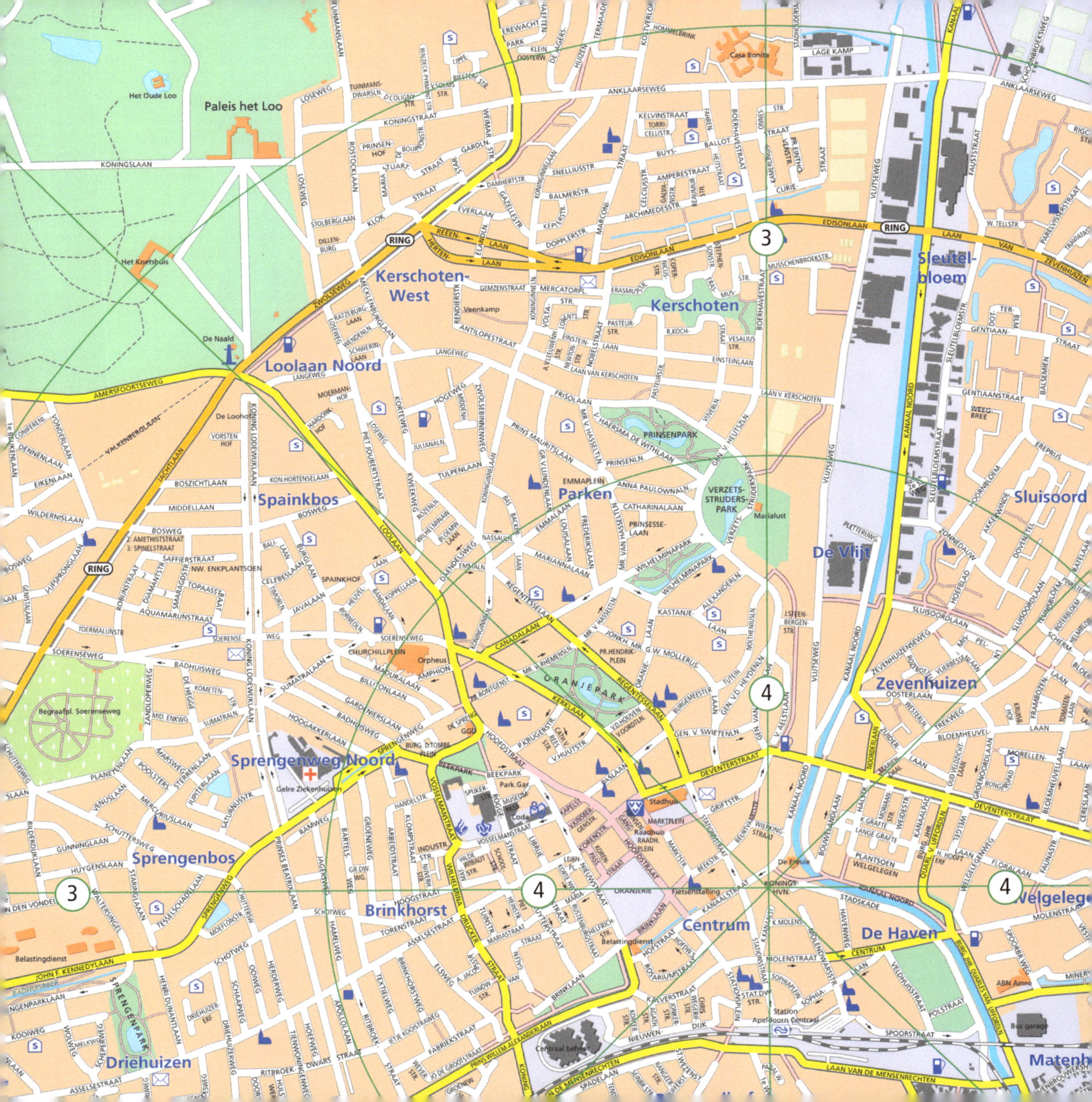

Paleis het Loo
Het Oude Loo
Het Koetshuis
De Naald
De Loohof
Casa Bonita
LAGE KAMP
Kerschoten-West
Kerschoten
Loolaan Noord
Sleutel-bloem
Spainkbos
Parken
PRINSENPARK
VERZETS-STRIJDERS-PARK
Marialust
De Vlijt
Sluisoord
Zevenhuizen
Sprengenweg Noord
Gelre Ziekenhuizen
Begraafpl. Soerenseweg
Sprengenbos
Brinkhorst
Centrum
De Haven
Welgelegen
Driehuizen
SPRENGENPARK
Belastingdienst
CHURCHILLPLEIN
Orpheus
ORANJEPARK
Stadhuis
MARKTPLEIN
Raadhuis
Park.Gar.
Coda
ORANJERIE
Fietsenstalling
De Freule
Station Apeldoorn Centraal
Centraal beheer
Bus garage
ABN Amro
Veenkamp
RING
3
4
KONINGSLAAN
ANKLAARSEWEG
KONINGSTRAAT
ZWOLSEWEG
AMERSFOORTSEWEG
JACHTLAAN
EDISONLAAN
LOOLAAN
DEVENTERSTRAAT
SOERENSEWEG
KANAAL NOORD
KANAAL ZUID
VLIJTSEWEG
LAAN VAN DE MENSENRECHTEN
JOHN F. KENNEDYLAAN
PRINS WILLEM-ALEXANDERLAAN
HOOFDSTRAAT
REGENTESSELAAN
CANADALAAN
WILHELMINAPARK
KERKLAAN
SPRENGENWEG
MOLENSTRAAT
STATIONSSTRAAT
Matenh

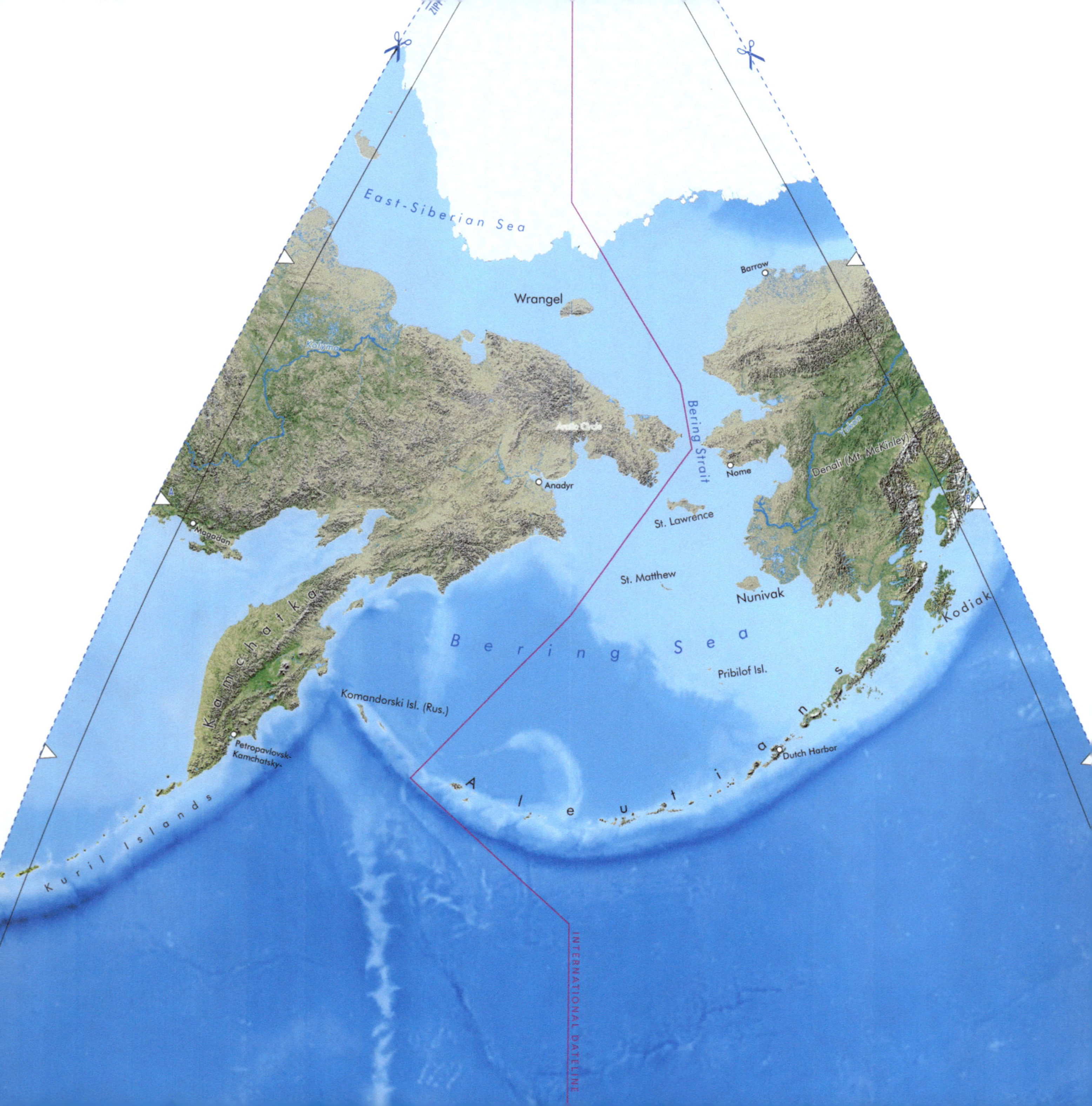

East-Siberian Sea
Wrangel
Barrow
Kolyma
Bering Strait
Arctic Circle
Anadyr
Nome
Yukon
Denali (Mt. McKinley)
St. Lawrence
Magadan
St. Matthew
Nunivak
Kamchatka
Kodiak
Bering Sea
Pribilof Isl.
Komandorski Isl. (Rus.)
Petropavlovsk-Kamchatsky-
Dutch Harbor
Aleutians
Kuril Islands
INTERNATIONAL DATELINE

Hans van der Maarel, Red Geographics

Oolaalaa Globe Chair

Software used: FME, Manifold System, Adobe Illustrator, Adobe Photoshop

More information: www.redgeographics.com

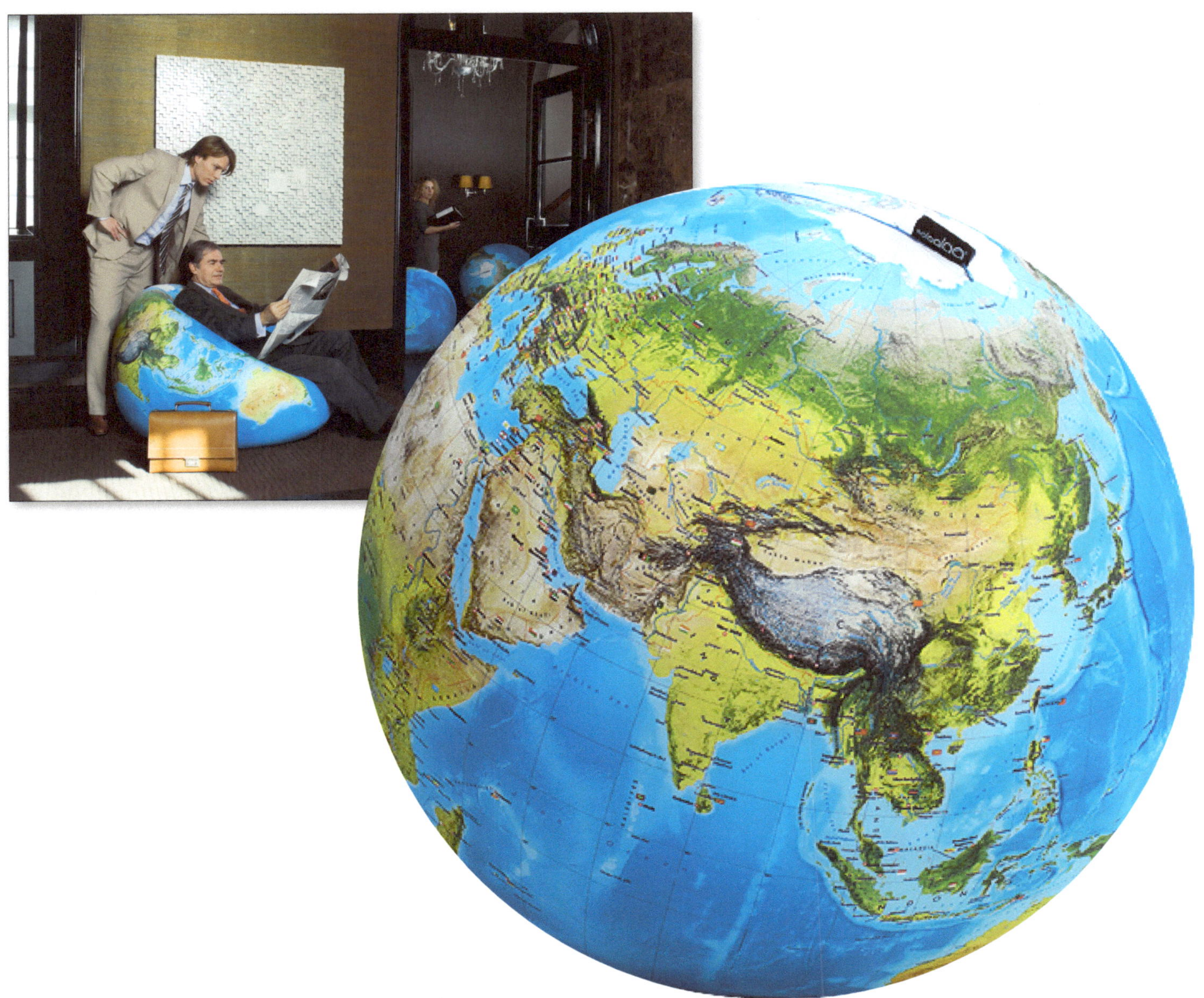

Martin von Wyss, vW Maps Pty Ltd

The Wine Map of Victoria by Max Allen

Map by Martin von Wyss, text by Max Allen

Software used: Manifold System, Adobe Illustrator, Adobe Photoshop

Data sources: VineFinders™, Commonwealth of Australia (Geoscience Australia), Australian Bureau of Meteorology, SRTM30

More information: www.vwmaps.com

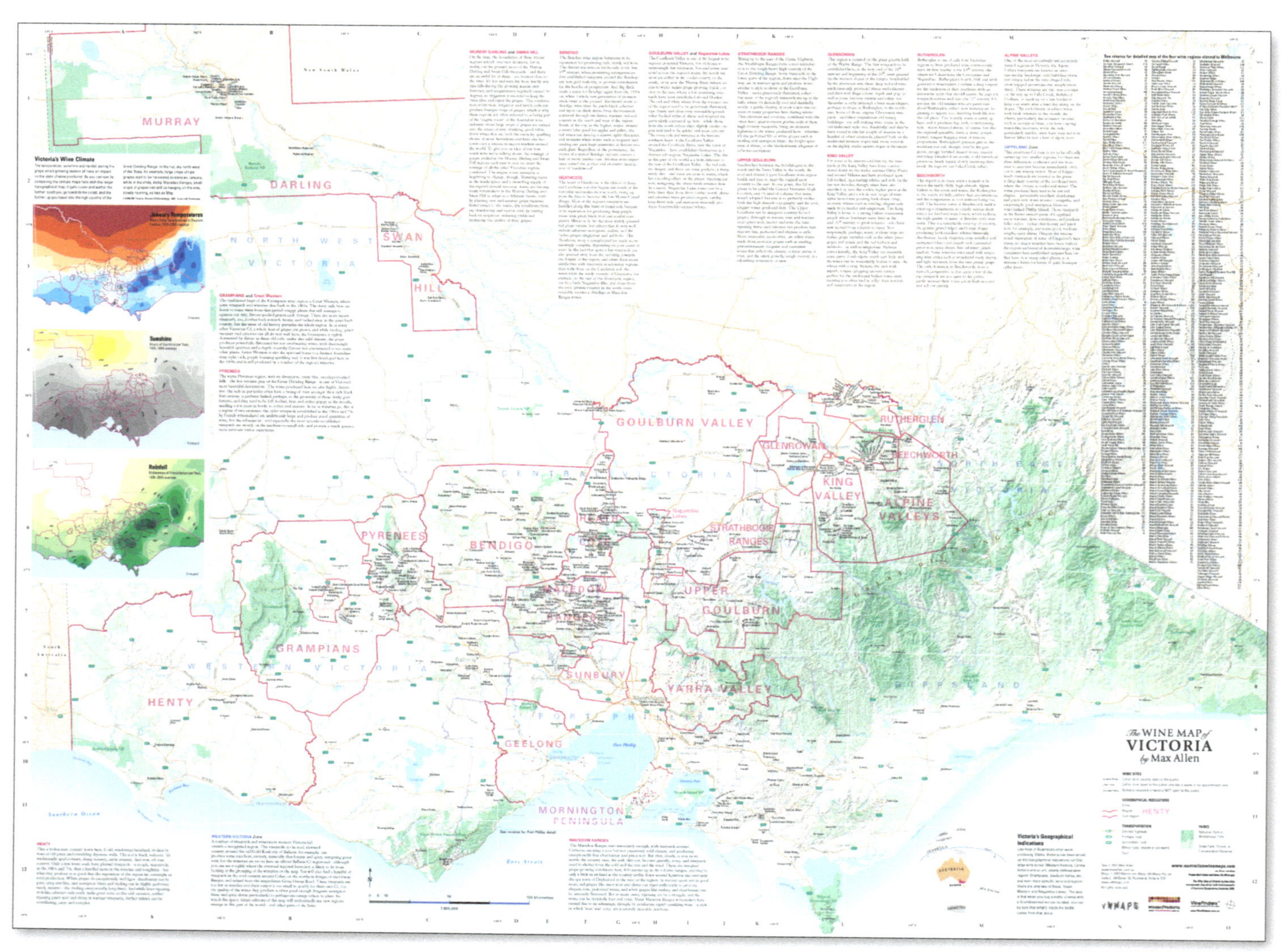

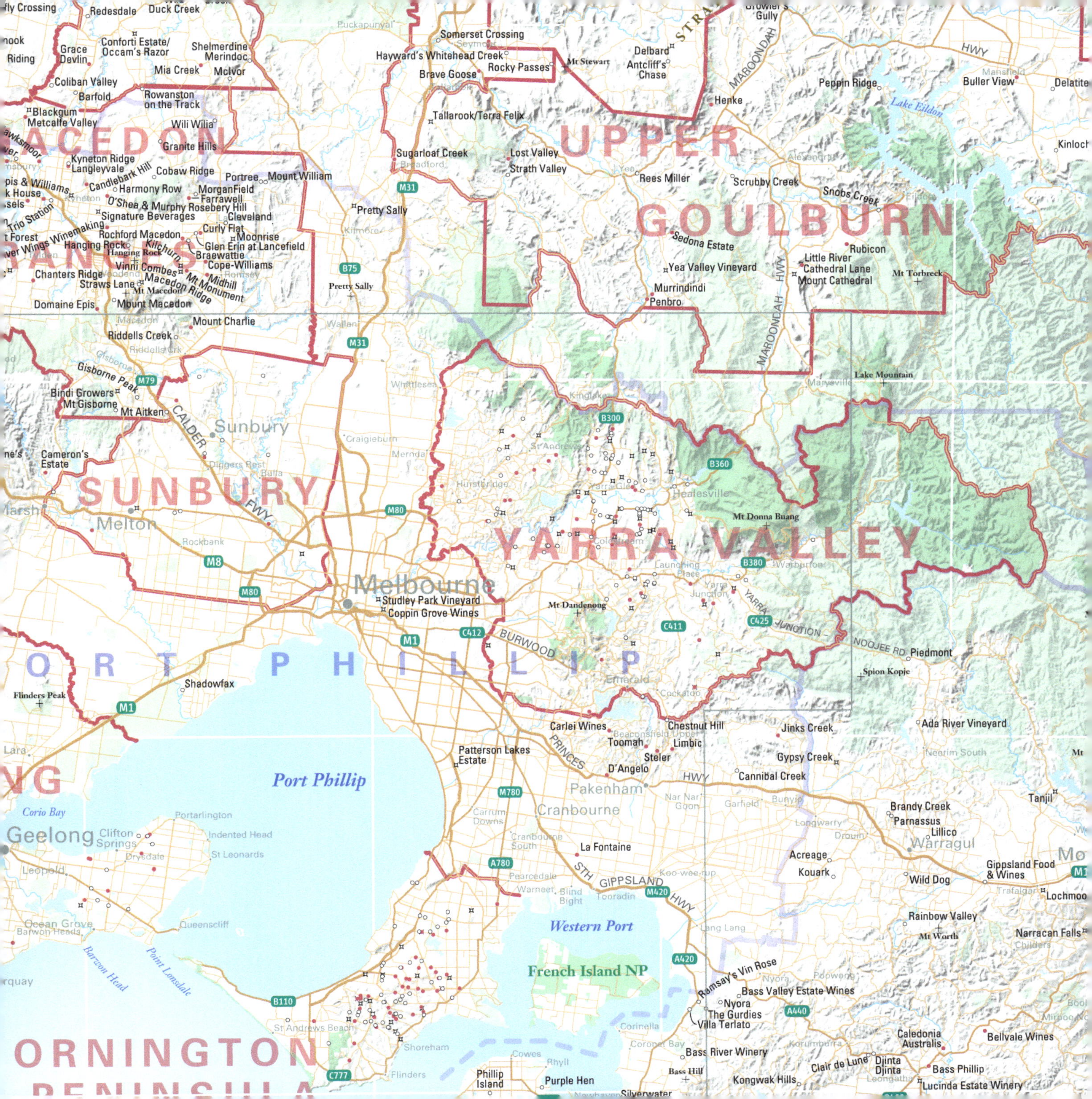

MACEDON RANGES
SUNBURY
UPPER GOULBURN
YARRA VALLEY
PORT PHILLIP
MORNINGTON PENINSULA
Melbourne
Port Phillip
Western Port
French Island NP
Geelong
Studley Park Vineyard
Coppin Grove Wines
Somerset Crossing
Hayward's Whitehead Creek
Rocky Passes
Brave Goose
Tallarook/Terra Felix
Sugarloaf Creek
Lost Valley
Strath Valley
Pretty Sally
Delbard
Antcliff's Chase
Henke
Peppin Ridge
Buller View
Delatite
Rees Miller
Scrubby Creek
Snobs Creek
Sedona Estate
Yea Valley Vineyard
Rubicon
Little River
Cathedral Lane
Mount Cathedral
Murrindindi
Penbro
Redesdale
Duck Creek
Conforti Estate/ Occam's Razor
Grace Devlin
Shelmerdine
Merindoc
Mia Creek
McIvor
Coliban Valley
Barfold
Rowanston on the Track
Blackgum
Metcalfe Valley
Wili Wilia
Granite Hills
Kyneton Ridge
Langleyvale
Candlebark Hill
Cobaw Ridge
Portree
Mount William
Harmony Row
MorganField
O'Shea & Murphy
Rosebery Hill
Farrawell
Signature Beverages
Cleveland
Curly Flat
Rochford Macedon
Moonrise
Glen Erin at Lancefield
Hanging Rock
Braewattie
Cope-Williams
Vinni Combes
Chanters Ridge
Straws Lane
Midhill
Domaine Epis
Mount Macedon
Mount Charlie
Riddells Creek
Gisborne Peak
Bindi Growers
Mt Gisborne
Mt Aitken
Cameron's Estate
Sunbury
Melton
Shadowfax
Flinders Peak
Carlei Wines
Toomah
Chestnut Hill
Limbic
Jinks Creek
Gypsy Creek
Steler
D'Angelo
Cannibal Creek
Ada River Vineyard
Piedmont
Spion Kopje
Patterson Lakes Estate
Pakenham
Cranbourne
La Fontaine
Brandy Creek
Parnassus
Lillico
Warragul
Tanjil
Acreage
Kouark
Gippsland Food & Wines
Wild Dog
Lochmoor
Rainbow Valley
Narracan Falls
Ramsay's Vin Rose
Bass Valley Estate Wines
Nyora
The Gurdies
Villa Terlato
Bass River Winery
Kongwak Hills
Clair de Lune
Djinta Djinta
Caledonia Australis
Bellvale Wines
Bass Phillip
Lucinda Estate Winery
Purple Hen
Phillip Island
Clifton Springs
Mt Dandenong
Mt Donna Buang
Healesville
Lake Mountain
Mt Torbreck
Lake Eildon
YARRA JUNCTION - NOOJEE RD
PRINCES HWY
STH GIPPSLAND HWY
CALDER FWY
MAROONDAH HWY
BURWOOD

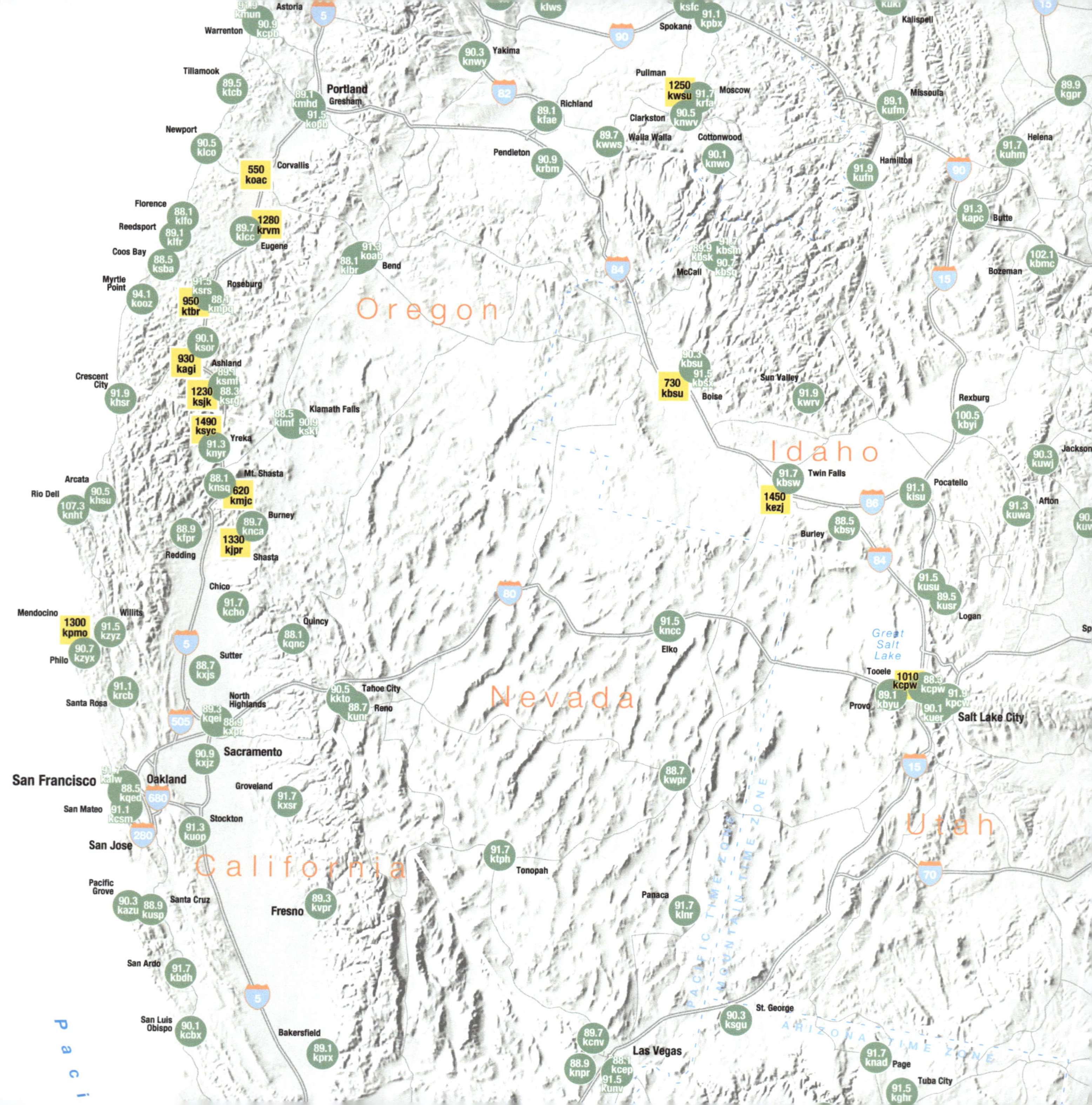

Oregon
Idaho
Nevada
California
Utah
Astoria
Warrenton
Tillamook
Portland
Gresham
Newport
Corvallis
Florence
Reedsport
Coos Bay
Eugene
Myrtle Point
Roseburg
Bend
Ashland
Crescent City
Klamath Falls
Yreka
Mt. Shasta
Arcata
Rio Dell
Burney
Redding
Shasta
Chico
Mendocino
Willits
Quincy
Philo
Sutter
Santa Rosa
North Highlands
Tahoe City
Reno
Sacramento
San Francisco
Oakland
Groveland
San Mateo
Stockton
San Jose
Pacific Grove
Santa Cruz
Fresno
San Ardo
San Luis Obispo
Bakersfield
Yakima
Spokane
Pullman
Moscow
Clarkston
Richland
Walla Walla
Cottonwood
Pendleton
McCall
Boise
Sun Valley
Twin Falls
Burley
Pocatello
Logan
Elko
Tonopah
Panaca
Las Vegas
St. George
Page
Tuba City
Kalispell
Missoula
Helena
Hamilton
Butte
Bozeman
Rexburg
Jackson
Afton
Great Salt Lake
Tooele
Provo
Salt Lake City
PACIFIC TIME ZONE
MOUNTAIN TIME ZONE
ARIZONA TIME ZONE
Paci

Martin von Wyss, vW Maps Inc.

The NPR Map

Software used: Manifold System, Adobe Illustrator, Adobe Photoshop

Data sources: NPR Member and Program Services, U.S.F.C.C., SRTM30, U.S.D.O.T.

More information: www.vwmaps.com

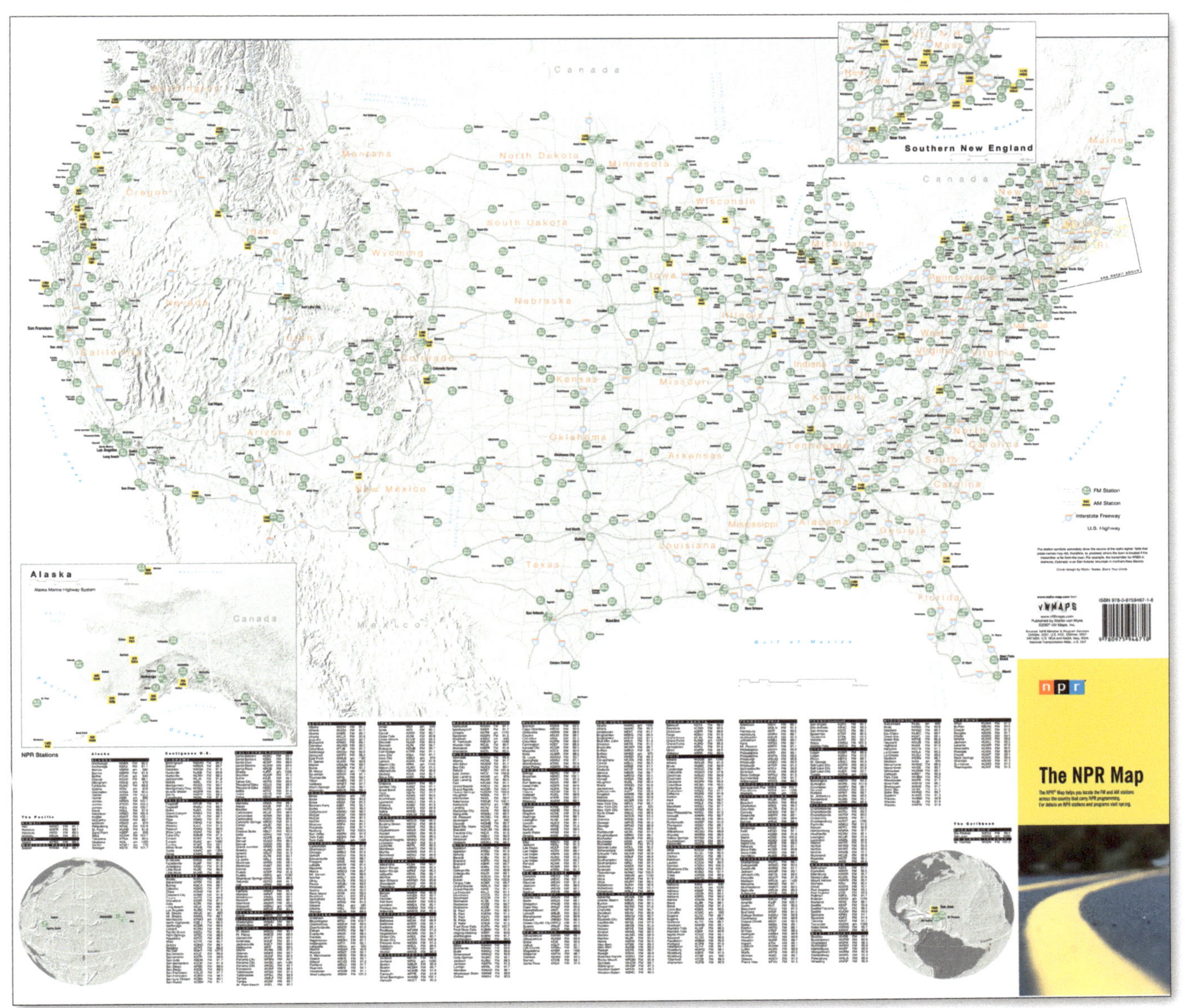

www.ingramcontent.com/pod-product-compliance
Lightning Source LLC
LaVergne TN
LVHW070141110826
845147LV00002B/302

* 9 7 8 0 6 1 5 2 2 1 1 6 8 *